Assessment in
SCIENCE

Practical Experiences and Education Research

Assessment in SCIENCE

Practical Experiences and Education Research

Edited by Maureen McMahon,
Patricia Simmons, Randy Sommers,
Diane DeBaets, and Frank Crawley

National Association for Research in Science Teaching

NATIONAL SCIENCE TEACHERS ASSOCIATION

Arlington, Virginia

NATIONAL SCIENCE TEACHERS ASSOCIATION

Claire Reinburg, Director
Judy Cusick, Senior Editor
Andrew Cocke, Associate Editor
Betty Smith, Associate Editor
Robin Allan, Book Acquisitions Coordinator

Will Thomas, Jr., Art Director
Tracey Shipley, Cover and Inside Design

PRINTING AND PRODUCTION Catherine Lorrain, Director
   Nguyet Tran, Assistant Production Manager
   Jack Parker, Electronic Prepress Technician

NATIONAL SCIENCE TEACHERS ASSOCIATION
Gerald F. Wheeler, Executive Director
David Beacom, Publisher

LIBRARY OF CONGRESS CATALOGING-IN-PUBLICATION DATA
Assessment in science : practical experiences and education research / Maureen McMahon ... [et al.],
editors.
   p. cm.
 ISBN-13: 978-1-933531-00-7
 ISBN-10: 1-933531-00-2
 1. Science--Study and teaching (Elementary)--Evaluation. 2. Science--Study and teaching
(Secondary)--Evaluation. 3. Science--Ability testing. I. McMahon, Maureen.
 LB1585.A785 2006
 507'.1--dc22
                          2006002237

contents

1 A Knowledge-Based Framework for the Classroom Assessment of Student Science Understanding

Recent research in cognitive science linked with established principles of educational assessment provide a framework that science teachers and researchers can use to assess student understanding of science. Knowledge-based categories for assessing science understanding were developed to reflect the major types of performance that expert scientists use to display their understanding of science concepts. These performance categories were integrated into a knowledge-based framework for educational assessment consisting of a general strategy for classroom assessment of science understanding.

2 Developing Assessment Items: A How-to Guide

Many current high-stakes accountability measures take the form of traditional test items. While traditional forms of assessment may not be the best method to assess all important skills and knowledge, teachers should learn how to construct these kinds of tests. Teachers can pose questions about their own teaching and student learning, seek answers in the form of student test results, evaluate those results, and then use that information to make instructional decisions. Helpful tips for generating and evaluating multiple-choice, short-answer, essay, true/false, and matching questions are reviewed.

3 Assessment in Support of Conceptual Understanding and Student Motivation to Learn Science

Classroom-based assessment strategies may influence the development of conceptual understanding and motivational beliefs among elementary learners in science. A contextual analysis of how young children (65 second graders) responded to classroom-based assessment—and the impact that assessment may have had on science learning—suggests that these young children enjoyed learning about science. Their positive attitudes may have developed because of the opportunities for active exploration they were given and the intellectual stimulation that resulted from new and unexpected discoveries in science. Successful assessment experiences may also have contributed to the enthusiasm that these students expressed for science.

"Adaptive inquiry" is the product of the synergistic relationship between what a student brings to the classroom, the teacher's ability to shape a lesson in response to the needs of the student, and the method of final assessment. This new form of assessment requires teachers to reexamine curriculum, instructional techniques, and measures of achievement. Effective teachers need to possess broad pedagogical knowledge, a content base, and a pedagogical-content repertoire that varies from direct instruction to open-ended inquiry in order to enhance student achievement.

School districts and individual schools must place a premium on linking classroom instruction, science content standards, and classroom assessment. In this study, the authors examined the influence of national science standards on middle school science teachers' classroom assessment practices. Middle school science teachers' classroom assessment practices were influenced by state science content standards and benchmarks and to a lesser degree the national science content standards and benchmarks.

This chapter reports on an approach to developing middle school science assessments using a learning-goals-driven design model. The design process for creating usable assessments that are aligned with curriculum and important science content and inquiry learning standards is described, as is the use of one assessment tool, rubrics. Evidence from the enactment of a middle school chemistry unit shows the initial success of the work reported on, as well as lessons learned from the real-world environment of an urban science classroom.

The Science Education for Public Understanding Program (SEPUP), located at the Lawrence Hall of Science (University of California-Berkeley) developed an assessment approach that included rubrics as well as other teacher tools. Middle school teachers incorporated this approach into their teaching and used authentic, embedded assessments linked to both the content and processes of science. Through the use of rubrics, students and teachers developed a clear sense of the expectations for success on a project and students were encouraged to focus on learning rather than grades.

Two urban middle school science teachers used concept maps as a form of assessment in their inquiry-oriented teaching. Critical steps in this learning process included adapting curriculum materials to a specific school context, adjusting and re-evaluating strategies, contextualizing concept mapping for students, and finding time and structures for peer collaboration. The two teachers in this study used professional development as an opportunity to interact with other teachers in similar teaching contexts with similar learning outcome goals.

The work discussed here focused on the design of curricula that mirrored important aspects of realistic scientific practice, especially the ways in which causal models are used to account for patterns in data and the use of criteria for judging the acceptability of explanations and/or knowledge claims. During a nine-week Earth-Moon-Sun (EMS) astronomy unit for ninth graders, students collected data, built models, and publicly justified their explanations. Students gained knowledge of EMS concepts and improved their inquiry skills and understanding of how knowledge is judged in science.

As science instruction moves from lectures that emphasize lists of facts to more student-centered approaches that emphasize knowledge generation and justification, it is clear that assessments must also change. The assessment instruments developed for a nine-week high school course on evolution engaged students in activities that mirrored the practices of evolutionary biology. Students used the model to explain phenomena through a set of writing assignments documenting their evolving ideas and allowing them to see how their ideas changed. By the end of the nine-week course, students made important gains in their understandings of variation, selection, and heritability.

New content standards call for the implementation of new and varied pedagogical interventions and instructional techniques. An experimental project in which chemistry was taught using new pedagogical and assessment standards with 10th- to 12th-grade students showed that students who participated in the project expressed satisfaction with the way they learned chemistry. This approach to assessment caused a significant decline in the student anxiety that accompanies the Israeli matriculation examinations. Teachers reported that participating in this project changed their teaching habits, moving them from delivering information to guiding students in the learning process.

A comprehensive assessment plan is a valuable component of a university's K–12 outreach effort. The K–12 Engineering Outreach Initiative has integrated an assessment plan into each stage of its outreach—development, implementation, and sustainability—with the ultimate goal of expanding the pool of youngsters who imagine and prepare themselves for futures in science, engineering, and technology. Outreach objectives were assessed by formative and summative assessment tools. The comprehensive assessment plan developed provides an analytical structure for guiding workshop development, shaping implementation, measuring success, and informing future planning. Lessons learned during this program are discussed.

The Performance Assessment project was a collaboration initiated by a university team of instructors and a local school district in which inservice and preservice teachers developed and implemented science performance assessment tasks during a one-semester science methods course. The development and administration of the performance assessment tasks required both inservice and preservice teachers to approach instruction in line with science education reform efforts. Both preservice and inservice teachers gained understandings and experiences in designing and implementing science performance assessment tasks.

If research into the assessment of science learning is going to make a difference for teachers, it must be applicable to the realities of K–12 classrooms. This research study conducted in urban elementary schools focused on how students and teachers engaged in both contextually authentic assessment and the new realities of high-stakes testing.

This collaboration between a fourth-grade teacher and university researcher resulted in the development of the Cognitive Strategies Inventory, an assessment instrument targeting writing, discussing, computing, reading, public speaking, and problem solving. The project focused on how an elementary teacher synthesized and applied four cognitive strategies (setting the purpose, relating prior knowledge, understanding metacognition, and looking for patterns) to classroom instructional activities.

The Classroom Assessment Project to Improve Teaching and Learning (CAPITAL), a collaborative research initiative between Stanford University and middle school science teachers in nearby school districts, examined classroom-based assessment in science. The teachers shared ideas with one another, and the university staff introduced research findings and ideas into the conversations. The CAPITAL teachers co-authoring this chapter described themselves as moving away from the role of teacher as giver of grades to teacher as conductor of learning. They increased their interactions with students during class time and better assisted students toward achieving the learning goals.

Many teachers are concerned about providing feedback to students in ways that support student learning. Five middle school teachers, part of the Classroom Assessment Project to Improve Teaching and Learning (CAPITAL), explored classroom-based assessment in science. The teachers experimented with different approaches to assessment and met regularly with other teachers and researchers to share and discuss their evolving practices. The CAPITAL teachers moved toward a more complex understanding of the nature of feedback, the purposes for feedback, and the roles feedback could play in their classrooms.

Mind mapping, a visual tool to improve note taking, foster creativity, organize thinking, and develop ideas and concepts, was the focus of this research project with middle school students. The efficacy of mind mapping as a teaching, learning, and assessment tool in a grade 6 science classroom was explored. This study provided preliminary support for the adoption of mind mapping as a flexible assessment tool to foster student learning in science and to guide curriculum planning and classroom practice.

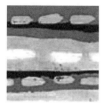

introduction

The National Science Teachers Association and the National Association for Research in Science Teaching have teamed up to create *Assessment in Science: Practical Experiences and Education Research*. This book, intended as a resource for teachers of science, teacher-researchers, and science education researchers, shares methods, stories, and findings about assessment, one of the most pressing issues in today's K–16 science classrooms. *Assessment in Science* links "best-practice" ideas to sound science education research.

Assessment, as defined in the National Science Education Standards (National Research Council 1996), is "a systematic, multi-step process involving the collection and interpretation of educational data" (p. 76). Planning appropriate assessment and evaluation is challenging—and even more difficult in the current climate of high-stakes testing.

The authors whose chapters appear in *Assessment in Science* are practicing K–16 classroom teachers or university-based educators and researchers. The following questions framed the call for papers to which these authors responded:

* Has your analysis of your students' work generated new strategies for your teaching?
* What are the toughest challenges in student assessment, and what are your best solutions?
* Has your research led to insights for identifying evidence of student understanding?
* What practical implications do research findings have for classroom, district, and national reform?

Each chapter is organized into an introduction or background section of relevant research, research questions or issues investigated, methods used to generate and analyze data, results from data analysis, and conclusions and implications for classroom practices. Most authors included a list of resources for further reading and summaries explaining how their studies aligned with the National Science Education Standards.

Authors of the first 13 chapters discuss how they: merged research findings from the cognitive sciences and assessment research base, conducted research on assessment, synthesized a new approach toward assessment for practitioners, and extrapolated research findings on assessment into highly practical information for teachers to employ in the classroom (such as constructing test items). Research studies were conducted at early elementary, elementary, middle, high school, and university levels. The authors employed qualitative and/or quantitative research methods to investigate their respective questions about assessment. Some studies were conducted on a large scale, while others focused on a select small number of individuals. The last 5 chapters contain reports of K–16 educators who were engaged actively in teacher-as-researcher roles. The results reported by all authors have implications for further research and for classroom practices to enhance the teaching and learning of science.

The authors present a variety of perspectives, from bridging theoretical backgrounds to practical tried-in-the-classroom tools:

* **Vitale**, **Romance**, and **Dolan** offer an overall knowledge-based framework for assessment and instruction. Synthesizing research from the cognitive sciences, they outlined a typology for assessing students' science understanding. Using the typology, they describe how teachers can generate student assessment activities and also meet assessment principles of test validity and reliability.

* **Henriques**, **Colburn**, and **Ritz** have organized their article as a helpful how-to guide, outlining tips and strategies for developing test items (multiple choice, true/false, matching, and essay) that align with instructional goals. They also included examples of ineffective and effective test items.

* **King** describes the results of a study with young children in early elementary grades where she functioned as a participant-observer. The students generally responded positively and enthusiastically to open-ended assessment opportunities.

* **Farenga**, **Joyce**, and **Ness** discuss how a state-mandated assessment required teachers to use strategies that are opposite to those required for science inquiry. Adaptive inquiry was the strategy they synthesized to explain how a middle school teacher dealt with this conundrum.

* **McWaters** and **Good** delve into how middle school teachers used national, state, and local science standards to guide their assessment practices. They recommend that teachers, administrators, and researchers collaborate more closely on projects involving classroom instruction, science content standards, and classroom assessment.

* **Harris**, **McNeill**, **Lizotte**, **Marx**, and **Krajcik** report on how they created assessments aligned with their curricular design process (learning-goals-driven

design) and the National Science Education Standards. They describe how specific rubrics were developed, implemented, and evaluated in a chemistry unit for middle grades students and teachers.

* **Siegel**, **Hynds**, **Siciliano**, and **Nagle** describe the use of authentic-embedded assessments (part of the SEPUP modules) that assisted students in creating a clear sense of expectations for learning and helped them focus on learning rather than grades. The authors provide examples for setting up rubrics, detail the thinking behind the generation of criteria for scoring guides, and report on teacher responses from field testing in middle schools.

* **Kubitsky**, **Fishman**, **Margerum-Leys**, **Fogleman**, **Brunvand**, and **Marx** discuss two urban, middle school teachers' implementation of concept mapping as one way to assess their students' learning.

* **Barton**, **Cartier**, and **Collins** describe a long-term collaboration between university researchers and high school science teachers that focused on curriculum design. They report on how teachers used multiple and varied assessments (in a ninth-grade astronomy unit) to determine gains in student knowledge, inquiry skills, and how knowledge is judged in science.

* **Passmore** and **Stewart** write about assessment instruments developed for a high school course in evolution. Investigating students' understanding and ability to use models to explain phenomena, they document the changes in students' ideas through specially designed writing assessment assignments.

* **Hofstein**, **Mamlok**, and **Rosenberg** describe the influence of a new high school chemistry project on changing teaching practices, teachers' beliefs, and students' views toward the new assessments. Both groups perceived that the project established healthy learning environments in their chemistry classrooms.

* **Knight** and **Sullivan** discuss the lessons learned from a K–12 university engineering outreach initiative. Their comprehensive assessment plan provided the analytical structure to guide project activities, measure success, and inform future planning on the initiative.

* **Akerson**, **McDuffie**, and **Morrison** describe the partnerships between preservice and inservice teachers during a science methods course that resulted in mutual learning benefits. Their collaboration documented greater understanding by both groups about the role of performance assessments linked to the National Science Education Standards.

* **Buxton** writes about how conducting research in urban middle schools led him to question if new approaches to teaching assessment (contextually authentic inquiry models) can be reconciled with the demands of high-stakes testing.

* **Flick** and **Tomlinson** describe how their professional collaboration (university/elementary educators) led to the implementation of four cognitive strategies in class. They recount how they synthesized the Cognitive Assessment Inventory, merging research on assessment from the reading comprehension and cognitive science literature bases.

* **Sato**, **Baker**, **Fong**, **Gilbertson**, **Liebig**, and **Schwartzfarb** explain how they (all but Sata are middle school teachers) moved away from the role of teacher as giver of grades to teacher as conductor of learning, through their participation in a university-school research collaboration (CAPITAL).

* **Cheung**, **Cody**, **Hirota**, **Rubin**, **Slavin**, **Coffey**, and **Moorthy** focus their article on five middle school science teachers (part of a larger research-practice-collaboration) who changed their beliefs and actions about assessment. The teachers' reflections form the central component of this chapter.

* **Goodnough** and **Long** describe a study in which mind mapping (a visual assessment tool) was used to examine middle school students' understanding of science and understanding of team work.

Readers are also referred to the websites listed below for additional, current information on K–16 assessment.

American Association for Higher Education
www.aahe.org/assessment/web.htm

Annenberg/CPB Project's Website for Learners and Educators: Assessment in Math and Science
www.learner.org

Buros Center for Testing
www.unl.edu/buros/

National Center for Research on Evaluation, Standards, and Student Testing
http://cresst96.cse.ucla.edu/index.htm

National Science Education Standards
www.nap.edu/readingroom/books/nses/html/

National Science Teachers Association (NSTA)—Position Statement on Assessment in Science Teaching
www.nsta.org/positionstatement&psid=40

No Child Left Behind: Regulations, Information, Implications
www.ed.gov/nclb/landing.jhtml?src=pb

Northwest Regional Educational Laboratory—Assessment
www.nwrel.org/assessment/

U.S. Departments of Education: State Comprehensive Testing, Accountability and Assessment—All 50 States
www.eduhound.com/k12statetesting.cfm

We thank the authors and Claire Reinburg and David Beacom at the National Science Teachers Association for their patience during the assembly of this volume. We extend our appreciation to the reviewers for their timely and helpful commentary and feedback. In addition, we thank Connie Quinlan (University of Missouri-St. Louis) for her assistance with the preparation of the articles.

about the editors

Maureen McMahon teaches elementary science methods at California State University, Long Beach, where she chairs the Department of Science Education.

Patricia Simmons holds the Orthwein Endowed Professorship in Life-Long Learning in the Sciences at the University of Missouri-St. Louis (UMSL) and is director of the Institute for Mathematics and Science Education and Learning Technologies.

Randy Sommers is completing his doctorate in educational leadership at UMSL. He is a former elementary school teacher in Missouri.

Diane DeBaets is a research assistant at the Institute for Mathematics and Science Education and Learning Technologies at UMSL.

Frank Crawley, chair of the Department of Science Education at East Carolina University, teaches courses in physical science, collaborative action research, and learning and teaching science.

a knowledge-based framework for the classroom assessment of student science understanding

Michael R. Vitale, Nancy R. Romance, and Megan F. Dolan

PRINCIPLES OF CLASSROOM ASSESSMENT

Four established principles of educational assessment can be applied to classroom practices (Mintzes, Wandersee, and Novak 1999; Pellegrino, Chudowsky, and Glaser 2001; Vitale and Romance 1995, 1999b). The first principle is that *a test (or assessment task) is considered a means for obtaining a sample of student behavior.* In this definition, student behavior is either anything that is observed directly (e.g., oral responses, student performance on a task) or the resulting record of student behavior that could have been observed (e.g., student written responses to test questions, a student-constructed project). The second principle is that *the purpose of obtaining the sample of student behavior through the test is to make an inference or conclusion about other student behaviors of primary interest that could be observed (or recorded) but that were not observed.* For example, if a teacher gave a student a set of 10 mathematics addition problems to assess the student's proficiency in addition, then the score on the 10 items would be used to make a conclusion or inference about the student's addition proficiency. That is, student performance on the 10 test items would be a means to an end—a conclusion or inference, not an end in itself. In this case, the resulting conclusion would be an inference on how the student might perform on all other possible addition problems.

All forms of classroom assessment reflect the two preceding assessment principles. A test is administered or an assessment task is assigned to make a conclusion about student proficiency. This proficiency takes the form of an inference about how a student would perform on a larger set of representative tasks. These principles are important because they provide a perspective for considering a third assessment principle, test validity: *Validity indicates the extent to which the test or assessment task that is administered is, in fact, credible for making the types of conclusions or inferences desired.* In a classroom, the concept of validity is based primarily on the extent to which the specific test items (or assessment tasks) represent the broader domain about which the conclusion or inference is made. Ultimately, the validity of any such test is established by demonstrating that it is an effective predictor of proficiency in some broad domain (Pellegrino, Chudowsky, and Glaser 2001; Vitale and Romance 1995). In this type of formal assessment (termed a *criterion-referenced assessment*), the issue of validity is addressed by developing domain specifications for constructing tests or test items. Such test specifications stipulate how the resulting items are representative of the domain itself.

The fourth assessment principle is reliability. In assessment, reliability commonly refers to *the degree of consistency of the measurement outcome (i.e., correct/incorrect score assigned).* If a test is reliable, then an individual student should obtain approximately the same score if the test were to be administered again. In assessment, reliability (like validity) is a matter of degree. A test may have high reliability, moderate reliability, or low reliability. If a test has high reliability, then the scores are considered very stable, and students would be expected to obtain approximately the same scores if the test were administered again. If a test has low reliability, however, the scores would be considered unstable, and student scores would be expected to vary greatly and randomly if retested. Typically, the greater the amount of subjective judgment, the less reliable the test scores will be. Subjective judgments may, however, have high reliability if judges are trained.

In assessment, low reliability is considered to reflect not only scoring judgments, but other factors, such as standardization of administration, randomly occurring factors during administration (e.g., room temperature), and individual, unrelated student experiences that could affect scores (e.g., headaches). Another important factor that affects reliability is the extent to which the specific items used in a test are representative of the broader content domain. For example, in the addition test, if two 10-item tests were constructed by sampling from all possible addition items, the extent to which student scores differed on the two tests would be an indication of reliability. For criterion-referenced tests, the concepts of curricular validity and test reliability are closely related through item-sampling issues (Crocker and Algina 1986; Vitale and Romance 1995). In considering the four assessment principles, it is important to note that they apply equally to all forms of test modalities (e.g., oral/written, multiple choice/essay, performance/portfolio).

A COGNITIVE SCIENCE VIEW OF MEANINGFUL SCIENCE UNDERSTANDING

Cognitive science is an interdisciplinary area of research that includes investigations of the role of knowledge in learning and performance (Vitale and Romance 2005). Recent applications of cognitive science to meaningful school learning (Bransford, Brown, and Cocking 2000; Mintzes, Wandersee, and Novak 1998;

Vosniadou 1996) reflect four common themes. First, for any content area, both the curriculum structure and student curriculum mastery of that content area should be consistent with the conceptual organization and proficiency of experts in that field. A cognitive science perspective of curriculum mastery considers mastery to be a form of expertise. This perspective is based on research that has consistently shown that both conceptual understanding and problem-solving applications of scientific knowledge by experts is substantially different from that of novices (Chi, Feltovich, and Glaser 1981). In terms of organizing and accessing their knowledge, experts are able to structure their in-depth conceptual understanding according to core concepts and core concept relationships that follow the logical structure of a discipline (Dufresne et al. 1992; Heller and Reif 1984; Leonard, Dufresne, and Mestre 1994). From the standpoint of cognitive science, the structure of any science curriculum should be based on the organization of the core concepts and concept relationships of experts in the sciences.

Second, research findings from cognitive science suggest that all classroom instruction should be organized according to curricular structure of core concepts and concept relationships (Bransford, Brown, and Cocking 2000). Just as classroom science instruction should focus on core science concepts, so should classroom assessment. This principle represents a *knowledge-based approach* to instruction (Vitale, Romance, and Dolan 2002) and always requires (1) an explicit representation of the knowledge to be taught and (2) an explicit linkage of all instructional and assessment activities to the core-concept curricular framework representing that knowledge. As part of a knowledge-based approach, both teachers and students relate the specific instructional or assessment activities to the overall core-concept curricular framework (Romance and Vitale 2001; Vitale and Romance 1999b). All classroom assessment should focus on student mastery or nonmastery of core science concepts or of subconcepts and examples related to core concepts.

Third, research findings from cognitive science suggest guidelines with implications for both classroom instruction and assessment (Bransford, Brown, and Cocking 2000; Vitale and Romance 1999b; Vitale, Romance, and Dolan 2002). Students must learn the core-concept framework that underlies instruction because it facilitates their cumulative development of in-depth understanding and parallels that of experts. This is accomplished by continually relating what students are asked to do in instruction and assessment to the core concepts and relationships that such activities represent. In working toward developing conceptual understanding, students must have a variety of learning (and cumulative review) experiences that are widely distributed across time (Grossen et al. 1998). Having a variety of experiences, and then relating these experiences to more general concepts, is a key to avoiding rote learning outcomes.

Fourth, in classroom learning and assessment, it is important to recognize the distinction between *declarative knowledge* and *procedural knowledge*. Declarative knowledge is what students know about a topic, while procedural knowledge is how students are able to apply such knowledge. In classroom settings, both forms of knowledge should be assessed.

Combining these implications from cognitive science research provides a comprehensive orientation for addressing in-depth understanding of science. Students should be able to use the conceptual organization they have gained to apply their knowledge to settings ranging from classroom experiments to real-life situations. In this chapter, we address how to approach the assessment of meaningful science understanding in classroom settings.

A Typology for Assessing Student Science Understanding

A typology for a knowledge-based approach to classroom assessment consists of distinct performance categories through which students demonstrate their understanding of science concepts (Romance and Vitale 2001; Vitale and Romance 1999b). The scope of the typology includes (1) declarative and procedural knowledge and (2) science conceptual content as well as an understanding of the nature of science. Classroom use of the typology requires specific instances of performance categories linked to the core-concept curricular framework (Vitale and Romance 1999b; Vitale, Romance, and Dolan 2002). Concepts are defined as classes or sets of instances that have one or more common characteristics. Scientific principles are represented as relationships among concepts. Analysis of observed events is based on knowledge of conceptual relationships.

The typology consists of six performance categories, which are based on how experts in science might organize and apply scientific knowledge. These categories represent how students are able to demonstrate meaningful science understanding as they gain expertise. These six categories are as follows:

Type 1: Observation or recording of events in the environment
Type 2: Prediction and/or manipulative control of future events
Type 3: Analysis and/or explanation of occurring events
Type 4: Construction of logical and/or visual representations (such as concept maps) of how science concepts are related to each other
Type 5: Demonstration of how science concepts and concept relationships are established (i.e., the nature of science)
Type 6: Communication of scientific knowledge to others

In general, the six categories follow a hierarchical structure, with an initial emphasis on concepts and concept relationships that are observable as the fundamental science knowledge to be learned. Given such fundamental knowledge, its application (Type 2, Type 3), representation (Type 4), expansion (Type 5), and communication (Type 6) must all be linked to the world (Type 1) if meaningful science understanding is to be accomplished. The first three categories (Types 1, 2, 3) represent scientific knowledge that provides a hierarchical foundation for the other three categories (Types 4, 5, 6).

To use the typology in assessment, it is important to be aware of methodological approaches for representing and using in-depth knowledge (Pellegrino, Chudowsky, and Glaser 2001; Romance et al. 1994; Vitale and Romance 1995, 1999b). One approach to knowledge representation in cognitive science uses *production rules*. These are constructed as IF/THEN statements, representing knowledge relationships among concepts. For example, the relationship *objects expand when heated* represented as a production rule would be: IF *an object is heated*, THEN *it will expand*. This knowledge provides students with the capacity to predict whether an object will expand or, by manipulating heat, to make an object expand (or contract). Production rules can also be used to represent knowledge of hierarchical classifications that are common to science (e.g., IF *a whale is a mammal* AND *mammals are warm-blooded*, THEN *whales are warm-blooded*). Finally, production rules are useful in representing the analysis and generation of possible explanations of events that have occurred (e.g., IF *bridge segments contract when cooled* AND *a specific bridge segment has contracted*, THEN *it is possible that the reason it contracted was that it cooled*). Because knowledge is represented in the form of conceptual relationships, IF/THEN production rules are powerful forms of representing scientific knowledge.

TOWARD A KNOWLEDGE-BASED FRAMEWORK FOR THE CLASSROOM ASSESSMENT OF SCIENCE UNDERSTANDING

Step-by-step guidelines and strategies in the form of a knowledge-based framework can help teachers to generate student assessment activities. It is important for teachers to recognize that being able to generate instructional and assessment activities within the same knowledge-based framework ensures that assessment of classroom instruction is valid.

The knowledge-based classroom science instructional and assessment framework consists of four steps.

1. *Teachers identify the core concepts and concept relationships to be taught by topic or unit.*
2. *Teachers construct a curriculum concept map representing the core concepts as a curriculum guide.* A simplified concept map for the concept of evaporation as an organizing guide for an instructional unit is shown in Figure 1.
3. *Teachers design activities or tasks for either instruction or assessment by combining selected elements from their concept map with appropriate categories from the assessment typology.* The curriculum concept map provides a structure for ensuring that the various activities used are representative of the scope of the unit (see Figure 1). Activities (or tasks) should be considered as different scenarios in which students are engaged (e.g., conducting hands-on experiments; observing teacher demonstrations; reading and discussing information in

FIGURE 1.

Simplified Illustration of a Propositional* Curriculum Concept Map Used By Grade 4 Science Teachers to Plan a Sequence of Science Instructional Activities

CURRICULUM CONCEPT MAP FOR FACTORS THAT AFFECT WATER EVAPORATION

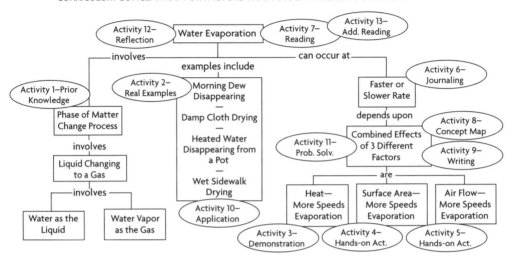

*The term propositional *refers to the fact that the links that join pairs of concepts are selected so that each concept-link-concept expresses a simple phrase in the form of a proposition (e.g., "Water evaporation can occur at faster or slower rate"; "Faster or slower rate [of water evaporation] depends upon combined effects of three different factors").*

Source: Romance, N. R., and M. R. Vitale. 2001. Implementing an in-depth expanded science model in elementary schools: Multi-year findings, research issues, and policy implications. International Journal of Science Education *23: 373–404. The figure is copyright © N. R. Romance and M. R. Vitale.*

textbooks, resource materials, or internet-based sources; writing or journaling; concept mapping key ideas and relationships). Each activity can be used as either an instructional or assessment activity. The transformation of activities into assessment assignments or test items is relatively easy to accomplish.

4. *Teachers connect all instructional and assessment tasks either to the initial teacher-developed core-concept map or to a concept map developed by students with teacher guidance.* This is a key element of the knowledge-based framework because it provides the means for relating specific instructional or assessment activities to the more general concepts and concept relationships they represent. In the context of assessment, the mapping of instructional activities and assessment activities onto curriculum concept maps ensures the validity of both in relation to the core concepts to be understood.

The primary, knowledge-based feature of this framework is the use of the core-concept map in conjunction with the assessment typology to generate activities or student tasks (Vitale 1998). All activities or tasks are explicitly knowledge-based. Once generated, they may be used with equal validity for instructional purposes by the teacher (i.e., as learning activities for students) or for student assessment. The same activity should not be used for both, because instructional activities may provide students with assistance.

STRATEGIES FOR GENERATING CLASSROOM ASSESSMENT ASSIGNMENTS AND TESTS
Teachers can use a series of strategies to construct classroom assessment assignments or tests from instructional activities.

Using classroom activities as assessment assignments

The simplest form of classroom assessment is to use activities that represent different categories of the performance typology. For example, students may be asked to make observations and record events (Type 1), make predictions about what would happen under specific conditions (Type 2), explain and justify why some event occurred (Type 3), create a concept map to illustrate concept relationships (Type 4), plan or conduct an experiment to test, explore, or establish relationships among concepts (i.e., scientific principles) (Type 5), and/or write a short report communicating any of the preceding items (Type 6). As long as student performance can be evaluated as successful or unsuccessful (i.e., using a rubric), and the activities are linked to and represent the curriculum core-concept framework, then such forms of assessment are valid measures of student science understanding. (We do concede that using activities for assessment may be highly inefficient in terms of time, scope of coverage, and use of classroom resources when compared to more standard forms of testing.)

Transforming activities and/or assessment assignments into tests

The term *test* in this section refers to the questions used to measure science understanding. These questions always refer to specific activities presented to the student in pictorial, graphic, and/or text form. The questions query the student about the activity represented or require the student to generate illustrative examples. For example, students could be asked to identify examples and nonexamples of a concept. Given a concept, students are asked to produce examples (Type 1); when presented with a specific set of conditions, asked to predict what will happen or, when presented with a desired outcome, asked to describe what actions will produce that outcome (Type 2); when presented with a description of some occurrence, asked to generate (and justify) plausible explanations regarding why it occurred (Type 3); when presented with a concept map with missing elements, asked to complete it (Type 4); when presented with a hypothesis that states the relationship among concepts, asked to design a study that would confirm or disconfirm that relationship (Type 5); and/or be asked to generate a short but elaborative report on a specific concept or concept relationships (Type 6). A clear advantage of using a standard test format is that teachers can use a wide variety of test questions and administer them efficiently to students. As long as the questions asked are linked to and represent the curriculum core-concept framework, then such questions would be valid measures of science understanding.

Teacher strategies for generating activities for assessment (or instruction)

Complementing the construction of curriculum concept maps (or the identification of core concepts) is the generation of activities for classroom use. A useful two-stage strategy for teachers is to:

1. Generate representative examples of the contexts (i.e., domains) in which the conceptual knowledge is applied by asking and answering the following question: "What different settings are representative of the scope of the concepts taught?"
2. Determine what student performance would provide evidence of conceptual understanding by asking and answering the following question: "In the settings identified, what would students (who understand the concepts) be able to do that would distinguish them from students who did not understand the concepts?"

By applying this two-stage strategy, teachers can develop informal specifications for student activities, adapt them for use as assessment questions (or tests), and then write a description of the assessment activity and question(s) that represents formal assessment specifications for the desired student performance in the following form:

* a description of the activity or test item,
* the outcome of the assessment activity or test item,
* the procedure or process to be followed by the student in completing the assessment activity or test item, and
* the conceptual knowledge needed by the student for completing the assessment activity or test item(s), referenced back to the curriculum concept map (making the activities and associated performance specifications knowledge-based).

It is important to note that the linkage of student performance specifications for the assessment activities or test items, student tasks and activities, and the curriculum concept map provides a framework through which teachers can organize and *refine both their course instruction and assessment* (Romance and Vitale 2001; Vitale 1998; Vitale and Romance 1995). The two-stage strategy can be used to design instruction that focuses on specified assessments (or test) specifications by working in a hierarchical fashion. Teachers can begin with the process and conceptual knowledge students need to perform successfully and then go back and identify prior conceptual knowledge needed to teach the new conceptual knowledge (see Figure 2 for examples of assessment assignments and test questions).

FIGURE 2.

Six Types of Performance Categories and Assessment Examples

Type 1: Observation or recording of events in the environment	
Assessment Assignment: The teacher provides several examples of water evaporating under various conditions and temperatures (e.g., water boiling in a pot, using a hot iron to dry a wet cloth, a paper towel drying in front of a fan blowing air). Students identify evidence of evaporation in each example.	*Test Question(s)*: Students are asked to correctly identify each of the following as examples or nonexamples of evaporation: drying off in the sun after a swim; ice melting; morning dew disappearing; clothes drying on a clothesline; water droplets appearing on the outside of a glass filled with a cold liquid.

Type 2: Prediction and/or manipulative control of future events	
Assessment Assignment: Students are shown wet paper towels that are spread out or crumpled and asked to predict which will dry more quickly, OR students are given wet crumpled paper towels and asked to dry the paper towels as fast as they can, manipulating whatever variables they wish.	*Test Question(s)*: Students are presented with a written scenario in which a crumpled paper towel and a spread-out paper towel are placed in front of a fan. Students are asked to predict which will dry more quickly, OR students are presented with a written scenario in which they are asked what they would do to make a crumpled paper towel dry more quickly.

Type 3: Analysis and/or explanation of occurring events	
Assessment Assignment: Several pans with different-sized openings (i.e., surface areas) are each filled with one cup of water and placed on a windowsill. Each hour throughout the day, students record how much water remains in each pan. At the end of the day, students are asked to explain why the water evaporated at different rates.	*Test Question(s)*: Students are presented with the following scenario: A shallow bowl is filled with one cup of water and placed on a shelf. A second identical bowl is also filled with one cup of water and placed in front of a fan. Students are asked to explain why the water in front of the fan evaporated more quickly.

Type 4: Construction of logical and/or visual representations (such as concept maps) of how science concepts are related to each other	
Assessment Assignment: Students are asked to read several sources about evaporation and construct their own concept maps, showing the concepts they have learned about water evaporation and how those concepts are related.	*Test Question(s)*: Students are presented with a map similar to Figure 1, but without the activities and with several empty boxes (concepts) and/or links (i.e., concept relationships). Students are asked to complete the concept map by writing in all missing information (i.e., empty boxes, missing links).

Type 5: Demonstration of how science concepts and concept relationships are established (i.e., the nature of science)	
Assessment Assignment: Given appropriate materials, students are asked to design and conduct an experiment investigating the effect of surface area (or other variables) on the rate of evaporation.	*Test Question(s)*: Students are asked to develop a plan for an experimental study that would investigate the effect of surface area (or other common variables) on the rate of evaporation.

(continued)

Type 6: Communication of scientific knowledge to others
Assessment Assignment: Students are asked to write a short report describing examples of evaporation they have observed; a description of an experiment demonstrating evaporation; an explanation of an everyday occurrence in terms of evaporation; a short explanation of evaporation using the concept map as a guide; or a research report of an experiment they conducted to investigate evaporation.	*Test Question(s)*: Students are presented with an informally written scenario (e.g., a newspaper-style report of a study) and asked to rewrite it using a more scientific format or style.

SUMMARY AND DISCUSSION

The overall knowledge-based framework (Vitale and Romance 1999a, 2001; Vitale et al. 2002) offers teachers an efficient, focused approach to both instruction and assessment. By developing and using a curriculum concept map (representing core concepts and concept relationships), teachers can ensure the validity of classroom activities for instruction and assessment. By following the two-stage strategy and using the curriculum concept map as a guide, teachers can generate multiple variations of classroom activities representing the content domain to be taught. In turn, using the six-part assessment typology, teachers can transform such classroom activities into assessment assignments or tests. Because these examples of assessment assignments or test items can be linked back to the organizational curriculum concept map, classroom assessment using such assignments or tests will have strong instructional validity (what is taught) and curricular validity (the science concepts to be understood). If teachers are careful in how assessments are conducted (e.g., they may be implemented and graded by others), then such assessments will also be reliable. Because these assessments are logical extensions of ongoing classroom observations of their students, teachers can meet reliability and validity standards in their classroom assessments. Finally, the knowledge-based framework provides a means for supporting teachers' revisions of their science courses to increase the in-depth science understanding by their students, either through individual or coordinated team efforts.

ACKNOWLEDGMENT

This work was supported by an Interagency Educational Research Initiative (IERI) grant from the National Science Foundation (Award Reference No. 0228353) and an Institute for Education Sciences (IES) grant from the U.S. Department of Education (Award Reference No. R305G040089).

LINKS TO THE NATIONAL SCIENCE EDUCATION STANDARDS

The National Science Education Standards (NSES) (NRC 1996) address the role and competencies of classroom teachers in a collegial sense as they support the broad vision of science education. In doing so, a set of teaching standards identifies the importance of assessment strategies that support the development of student understanding and nurture a community of science learners. Standards A and C (Science Teaching Standards, Chapter 3) address assessment as both a means to evaluate student performance outcomes and a source of data upon which decisions regarding curriculum and instruction can be made. As such, the vision supported by the Standards is one in which classroom teachers use student data, observations, and interactions with colleagues to reflect on and improve teaching practice. Further detail about the quality and characteristics of assessment practices used by teachers and state and federal agencies to measure student achievement are represented in another set of Standards (i.e., Assessment in Science Education Standards A–E, Chapter 5).

REFERENCES

Bransford, J. D., A. L. Brown, and R. R. Cocking. 2000. *How people learn: Brain, mind, experience, and school.* Washington, DC: National Academy Press.

Chi, M. T. H., P. J. Feltovich, and R. Glaser. 1981. Categorization and representation of physics problems by experts and novices. *Cognitive Science* 5: 121–152.

Crocker, L., and J. Algina. 1986. *Introduction to classical and modern test theory.* New York: Harcourt Brace Jovanovich.

Dufresne, R. J., W. J. Gerance, P. Hardiman, and J. P. Mestre. 1992. Constraining novices to perform expert-like problem analyses: Effects of schema acquisition. *The Journal of Learning Sciences* 2 (3): 307–331.

Grossen, B. J., D. W. Carnine, N. R. Romance, and M. R. Vitale. 1998. Effective strategies for teaching science. In *Effective teaching strategies that accommodate diverse learners,* eds. E. J. Kameenui and D. W. Carnine, 113–137. Upper Saddle River, NJ: Prentice Hall.

Heller, J. I., and F. Reif. 1984. Prescribing effective human problem solving processes: Problem description in physics. *Cognition and Instruction* 1: 177–216.

Leonard, W. J., R. J. Dufresne, and J. P. Mestre. 1994. Using qualitative problem solving strategies to highlight the role of conceptual knowledge in solving problems. *American Journal of Physics* 64: 1495–1503.

Mintzes, J. J., J. H. Wandersee, and J. D. Novak. 1998. *Teaching science for understanding: A human constructivist view.* Englewood Cliffs, NJ: Academic Press.

Mintzes, J. J., J. H. Wandersee, and J. D. Novak. 1999. *Assessing science understanding: A human constructivist view.* San Diego, CA: Academic Press.

Pellegrino, J. W., N. Chudowsky, and R. Glaser, eds. 2001. *Knowing what students know: The science and design of educational assessment.* Washington, DC: National Academy Press.

Romance, N. R., and M. R. Vitale. 2001. Implementing an in-depth expanded science model in elementary schools: Multi-year findings, research issues, and policy implications. *International Journal of Science Education* 23: 373–404.

Romance, N. R., M. R. Vitale, P. Widergren, and H. Parke. 1994. Developing science conceptual understanding through knowledge-based teaching: Implications for research. Paper presented at the Annual Meeting of the National Association for Research in Science Teaching, Anaheim, CA (April).

Vitale, M. R. 1998. *Assessment in college teaching: A restrictive instructional systems perspective.* Greenville, NC:

East Carolina University, Educational Research Laboratory.

Vitale, M. R., and N. R. Romance. 1995. Technology-based assessment in science: Issues underlying teacher advocacy of testing policy. *Journal of Science Education and Technology* 5 (2): 35–44.

Vitale, M. R., and N. R. Romance. 1999a. *A knowledge-based approach to content area comprehension.* Boca Raton, FL: Florida Atlantic University, Region V Area Center for Educational Enhancement.

Vitale, M. R., and N. R. Romance. 1999b. Portfolios in science assessment: A knowledge-based model for classroom practice. In *Assessing science understanding: A human constructivist view*, eds. J. J. Mintzes, J. H. Wandersee, and J. D. Novak, 168–197. San Diego, CA: Academic Press.

Vitale, M. R., and N. R. Romance, 2005. Research in science education: An interdisciplinary perspective. In *Teaching science in the 21st century*, eds. J. Rhoton and P. Shane, 329–351. Arlington, VA: NSTA Press.

Vitale, M. R., N. R. Romance, and M. Dolan. 2002. A rationale for improving school reform by expanding time for science teaching: Implications and opportunities for changing curricular policy and practice in elementary schools. Paper presented at the Annual Meeting of the National Association for Research in Science Teaching, New Orleans, LA (April).

Vosniadou, S. 1996. Learning environments for representational growth and cognitive science. In *International perspectives on the design of technology-supported learning environment*, eds. S. Vosniadou, E. DeCorte, R. Glaser, and H. Mandl, 13–24. Mahwah, NJ: Lawrence Erlbaum.

RESOURCES

General

Bransford, J. D., A. L. Brown, and R. R. Cocking. 2000. *How people learn: Brain, mind, experience, and school.* Washington, DC: National Academy Press.

> A very readable overview of research in cognitive science that focuses on meaningful conceptual learning, directly relevant to science instruction and assessment.

Pellegrino, J. W., N. Chudowsky, and R. Glaser, eds. 2001. *Knowing what students know: The science and design of educational assessment.* Washington, DC: National Academy Press.

> A comprehensive, nontechnical overview of concepts, perspectives, and recent developments in educational assessment.

Vitale, M. R., and N. R. Romance. 1999. Portfolios in science assessment: A knowledge-based model for classroom practice. In *Assessing science understanding: A human constructivist view*, eds. J. J. Mintzes, J. H. Wandersee, and J. D. Novak, 168–197. San Diego, CA: Academic Press.

> A brief introduction to elements of knowledge-based instructional models and how the representation and manipulation of conceptual knowledge serves as a framework for assessing science understanding.

Science Standards

National Research Council. 1996. *National science education standards.* Washington, DC: National Academy Press.

> Chapter 5 ("Assessment in Science Education") addresses the issue of assessment within a context of national standards for science education.

American Association for the Advancement of Science (in preparation). Resources for science and mathematics literacy: Assessment. *www.project2061.org/newsinfo/press/rl052899.html*

> This resource focuses on new strategies for the alignment of assessment with national, state, and local standards in science and mathematics and on guidelines for creating and improving standards-based assessment tasks.

AUTHOR AFFILIATIONS

Michael R. Vitale is a professor of curriculum and research in the College of Education at East Carolina University and co-principal investigator (with Nancy Romance) of a National Science Foundation (NSF) Interagency Educational Research Initiative (IERI) studying the process of scale-up of Science IDEAS, a grades 3–8 model for teaching comprehension through in-depth science instruction.

Nancy R. Romance is a professor of science education at Florida Atlantic University and the principal investigator of the NSF/IERI Science IDEAS scale-up project. She is also part of a U.S. Department of Education Institute of Educational Sciences project studying the effects of integrating reading comprehension strategies into science-oriented versus narrative-oriented learning environments in grades 3–5.

Megan F. Dolan is senior project coordinator for the Region V Area Center for Educational Enhancement at Florida Atlantic University. Her research interests are in the areas of curriculum development, assessment, and school improvement.

/

developing assessment items:
a how-to guide

Laura Henriques, Alan Colburn, and William C. Ritz

INTRODUCTION

When asked about assessment, most people think of tests or grades. We focus on "traditional" test items in this chapter because a large percentage of the high-stakes accountability measures use this format, and many science teachers have had little or no formal training in test construction. Although traditional forms of assessment may not be the best method for assessing all science skills and knowledge, teachers should learn how to construct these kinds of test items.

Assessment is much more than simply testing and grading; it is an ongoing process. Teachers pose questions and seek answers about student learning in the form of student performance on tests, in class activities, and in class discussions. These all serve as data with which teachers can make informed instructional decisions. In too many cases, teachers pose questions and collect data, but do not use that data to inform instructional or curricular decision making. This failure to "close the loop" can prevent teachers and students from profiting from assessment. The most well-constructed test items are of little value if they do not give information to teachers that they can use to inform instruction.

DEVELOPING TEST ITEMS

There are many different ways teachers can evaluate student knowledge. This section includes a description of these various ways, including methods for creating valid and appropriate items, and selected examples. Prior to creating any test item, teachers must think about their goals for students and the teaching strategies used to reach those goals. These goals not only influence how one teaches, but also help determine which assessment methods are most appropriate.

The classic forms of assessment—multiple choice, true/false, short answer, and matching—are viewed by students as being "objective." Students see that there is usually one "right" answer and view these assessment methods as good for assessing content knowledge and factual information. However, no one testing approach will work for all curricular and instructional goals. Following is a description of multiple-choice, short-answer, and essay items, among the most commonly used kinds of questions on tests. (In spite of the fact that students consider essay questions to be subjective and multiple-choice and true/false items to be objective, *all* test items are subjective because the teacher is the one who decides which items to include and which content to assess.) Information about true/false items and matching items can be found in the Appendix, pp. 28–29.

MULTIPLE-CHOICE ITEMS

All assessment options have advantages and disadvantages. Multiple-choice items allow teachers to ask students many questions in a short period of time. Teachers can sample and assess a wider range of topics than if they were to use matching, short answer, or essay questions, leading to more reliable test scores. Multiple-choice items allow for this wide assessment because they demand less reading and writing. Note that true/false items require even less reading time, meaning that more true/false items can be asked than multiple choice in the same amount of time. Well-constructed incorrect answers can provide important diagnostic information to teachers about student thinking and understanding of science concepts. Scoring is relatively easy and objective (note that objective *scoring* does not mean the overall *question* is objective).

Limitations of multiple-choice items include the difficulty associated with constructing valid items, a time-consuming practice. It is often challenging to write plausible distractors. Scores are influenced by students' reading ability (test scores correlate more closely to reading ability than scientific understanding). Multiple-choice items cannot be used to measure all instructional goals for students (i.e., problem solving, ability to communicate and express ideas, organizing data). Lastly, scores can be influenced by guessing (approximately 25% of the time students can guess the correct answer).

The question or partial statement of a multiple-choice test item is called the *stem*, and the answer choices are referred to as *foils*. Incorrect answers are known as *distractors*. Well-constructed multiple-choice test items are difficult to write. There are, however, some general rules to follow that will assist in writing valid multiple-choice items (see Figure 1 for sample *inappropriate* multiple-choice items):

1. Formulate a useful proposition or statement.

- Be sure that the stem contains only a single question or idea.
- Avoid using negative words. If using a negative word, be sure to underline or capitalize it so it stands out as students read the item.
- Minimize the reading demands on students.
- Avoid adding extraneous information to the question, unless you are testing whether or not students can distill important information.
- State the stem in clear, simple language.
- Avoid providing cues to the correct answer in the stem.

2. Translate the proposition into a question.
3. Develop a concise answer.
4. Develop plausible distractors.
   - Include the computational or conceptual errors that students regularly make as distractors. This provides attractive alternatives for students, and information about students' thinking when they select a particular distracter.
   - Use "all of the above" and "none of the above" sparingly. If these are used, be sure that they are occasionally the correct answer (otherwise, students will see them as "filler," and discard the distractors). "None of the above" is better to use since students must know that none of the answers are correct. "All of the above" is less discriminating, since students may realize that two choices are true but have no idea about the third. By default, "all of the above" has to be the correct answer.
   - Try not to use paired opposites in the choices (they imply one is correct and the other is incorrect; students will eliminate the other choices). *Homozygous* and *heterozygous* are paired opposites, *male* and *female* are paired opposites. Everything falls into one of the two categories, resulting in advantages to test-wise students.
   - Use common phrases and textbook language consistently.
   - Be cautious about distractors that converge to the right answer.
   - Write all choices approximately the same length.
   - Be careful with grammatical cues (e.g., asking for something in the plural but having distractors that are singular).

SHORT-ANSWER ITEMS

Short-answer questions include fill-in-the-blank or completion items and questions that require a word, phrase, date, number, or a few sentences to answer. These items differ from multiple choice, true/false, and matching items in that students are required to supply the answer rather than select an answer. Even though these questions require only brief responses, students view them as ambiguous, open to interpretation, and, therefore, subjective. An advantage of short-answer items is the relative ease with which teachers can write them. Also, since students supply an answer rather than selecting one, guessing is less likely to influence scores. Short-answer items are particularly useful for computational problems and assessing a wide

FIGURE 1. ▨

Sample Inappropriate Multiple-Choice Items

Below are examples of multiple-choice items that are inappropriate to use. The correct answer is in bold type with the critique of the item on the right.

Minor differences in organisms of the same kind are known as a. heredity. **b. variations.** c. adaptation. d. natural selection.	There are two problems with this question. First, the question asks for differences and choice (b) is the only answer that is plural. Second, the stem asks about differences. *Variations* is a synonym for *differences*. A student does not need to know about variations within organisms to correctly answer this item.
How long would it take a truck to increase its speed from 10 m/s to 30 m/s if it does so with uniform acceleration over a distance of 80 meters? a. 2.0 seconds b. 5.0 seconds c. 4.0 meters/second^2 **d. 4.0 seconds**	The foils converge to a correct answer. We know that (c) cannot be correct because the question asks for a time and the units in (c) are not time units. Both (c) and (d) have 4.0 as the numerical part of the answer and (d) has the proper units. The test-wise student will choose (d) without solving the problem.
In an experiment using fruit flies, a light bodied parent is crossed with a dark bodied parent. The offspring were all light bodied. Two light bodied offspring were then crossed, producing both light and dark bodied offspring in a ratio of 3 light to 1 dark. Using this information answer the following questions:	The use of a scenario for multiple-choice questions is good—it limits the reading demands.
The parents of the second (P_2) cross are **a. hybrid** b. pure c. homozygous for light body d. none of the above	"None of the above" is not a valid distractor. The foils include mutually exclusive subsets—hybrid and pure. As a result, the answer cannot be "none of the above."
The F_1 generation in the second cross is a. hybrid **b. pure** c. heterozygous for light body d. none of the above	These items contain cues in the answers. If a fly is homozygous it is pure. If a fly is heterozygous it is hybrid. By default the student can select the correct answer.
A seismic sea wave is properly called a a. sushi b. tidal wave **c. tsunami** d. tsumeb	Use distractors wisely! Do not give "throwaway" answers. *Tsumeb* is a made-up word, and most students will recognize that sushi is food. This narrows the choices to two. Any wave could be a tidal wave (a wave from the tides). *Tsunami* is a technical term that students will recognize as the answer.

Which of the following are known for their contributions to our understanding of electricity? a. Hans Christian Oersted and Isaac Newton b. Galileo Galilei and Thomas Edison c. Benjamin Franklin and Aristotle **d. Alessandro Volta and L. Galvini**	This item is not very effective because it discriminates against a student who is very knowledgeable. For example, Newton completed some work with electricity, even though this was not his primary area of work. Furthermore, the correct answer does not contain names of scientists that are most associated with the topic of electricity.
Which of the following is a characteristic of chlorophyll? a. Chlorophyll in plants is colorless. b. Chlorophyll is only found in plants. **c. Chlorophyll is integral to a plant's ability to carry out photosynthesis, converting water and carbon dioxide into sugar**. d. Chlorophyll causes diseases.	Length and vocabulary are both clues to this answer. The textbook-sounding answer of (c) makes it stand out from the others, as does its length. This item would be better if all foils were approximately the same length and used similar language.
Which of the following organisms has many legs? a. butterfly b. dog c. ant **d. spider**	Vague wording—all organisms could have many legs when compared to people! This question could be improved by asking which organism has the most legs.

range of student knowledge. Disadvantages of this assessment method are tied to issues of ambiguity. It is challenging to phrase statements so that a single answer is correct. Scoring is influenced by spelling and grammatical abilities and is more time consuming than multiple-choice, true/false, or matching items. Following are tips for short-answer items (see Figure 2 for sample *inappropriate* items):

* Make sure that the items cover important content, not trivia. In drafting items, ask yourself if a piece of information is crucial to understanding the scientific concepts studied.
* Be sure that the question or statement addresses a specific problem. Short-answer items can be vague or too broad, lending themselves to multiple correct responses.
* Make sure that the language used in the question is precise and accurate. Science has a specific vocabulary, which must be used correctly in questions.
* If a question requires numerical responses, it is helpful to indicate the units (e.g., minutes, centimeters) that students should use in their answers (helping with scoring and reducing ambiguity for students).
* Only omit key words in a completion item.
* Avoid using too many blanks in a single statement.
* Place/position blanks near the end of the statement instead of the beginning of the statement.

FIGURE 2. ▓▓▓

Sample Inappropriate Short-Answer Items

Below are examples of short-answer items that break the rules discussed on page 19. The critique of the item is on the right.

The names of two rivers that meet in Cairo, Illinois, are the _____ and _____.	The lengths of the blanks give away the answers. You should also have only one question/blank per item.
Thunderstorms form when columns of _____ air rise to cooler altitudes.	Verbal clue—the fact that air rises to cooler altitudes means that it must have been hotter first.
_____is the name given to represent an object's tendency to stay at rest.	The blank should appear at the end of the statement.
A scientist noted for his work with electricity is _____ ____.	This question is vague, because there are many plausible answers.

PERFORMANCE TASKS AND ESSAY ITEMS

Performance tasks require a product (such as essays, projects, lab reports, oral presentations, portfolio entries, journals, and on-demand tasks). Teachers assess performance tasks through some sort of systematic observation of either the process or the product. As a result, performance tasks tend to be more open-ended than multiple choice, true/false, or short answer and are often viewed as being more authentic or representative of real life. The scoring of performance tasks is also seen as more subjective than other ways of assessing, because students can complete the task in many different ways.

A key aspect to think about when constructing prompts for performance tasks is to consider what will be assessed—the process (you will be observing) or the product (you will review the finished work). Scoring performance tasks is very time consuming. If students supply long written answers, handwriting, spelling, writing skills, and "bluffing" all contribute to difficulties for deriving accurate scores. On the other hand, performance tasks are effective for measuring the highest levels of learning outcomes and can be used for complex tasks. Preparation time is less than that required for selection-type items. The key attribute is that real-world situations and the integration and application of ideas can be emphasized in this kind of assessment. Following are tips for writing essay items (these tips will also guide construction for other types of performance tasks):

* Try to use more essay questions that require briefer responses (rather than fewer essays requiring lengthier responses). This will help with sampling error. The greater the number of items (matched to goals), the more accurate the student scores.
* Write an "ideal" answer before using the item. Writing the response as part of test development will help focus the question. It will also indicate if an item is sufficiently specific for students to respond appropriately.
* Place physical separations between parts of an essay question. If there are multiple parts to a question, many students will invariably miss some of them. Using a bulleted or numbered list helps students organize their thoughts and ensures that they will answer all parts of the essay question.
* Be specific in wording questions, taking care to avoid vague terminology. Starting the question with phrases such as *compare, contrast, give the reasons for,* or *predict what would happen if* is much more specific (and will better guide students) than using *discuss.*
* Questions that are value laden should be judged based on evidence provided by the student, not the position taken by the student.
* Require all students to answer the same questions. Allowing students to choose which sections or questions to answer makes comparisons between student responses difficult to justify.
* Specify the number of points the essay is worth or the amount of time you expect students to write on the question. This helps students budget time and decide how to allocate and spend their efforts.

SCORING RUBRICS

Scoring rubrics, or scoring guides, are used to evaluate all types of performance tasks. Rubrics contain descriptions delineating levels of quality in responses. The essential characteristics of the desired response are included in rubrics. Teachers and others (including students) using the rubric should be able to reach the same conclusions about the quality of responses.

The teacher should develop rubrics before giving the assessment. Having the rubric in advance ensures that the question or prompt is specific. In performance assessments, it is often suggested that students be provided the rubric in advance. It is important to communicate with students and let them know what it is that the teacher wants them to know and be able to do. The rubric specifies what the teacher values, taking away guesswork. Teachers' goals and objectives should not be a mystery to students; likewise, how students are to be evaluated should not be a mystery or a surprise to them.

Teachers use rubrics to describe levels of quality of student work. The dimension or components of the expected response(s) are taken from the question prompt. For example, there should be a direct correlation between the prompts of an essay question and the components that students will be expected to display in their responses. Extraneous criteria should not be introduced.

The easiest way to create a rubric is to make a list of characteristics or criteria that will be the basis for judging the appropriate response(s). It is important for teachers to review the list and think about the best possible answer/product. What would it look like? How would the teacher describe it in terms of the criteria listed? A descriptor of the lowest level of work expected should be written next. This descriptor should include each of the criteria on the original list, differentiating the quality of work for each criterion. To create the middle-level descriptors, review the highest and lowest quality descriptors. What would be a logical midpoint for each of these descriptors? This midpoint is the middle-level score. (See a rubric for essays questions in Figure 3.)

Once the rubric is created, teachers should test it using samples of student work. Often rubrics need to be adjusted or have clarifications added to the descriptors. After scoring samples of student work, teachers can add examples to each of the descriptors based on actual work, making the performance-level descriptions more concrete.

The Rubistar website (*http://rubistar.4teachers.org/index.php*) demonstrates rubrics for various kinds of projects and assignments. The pre-made rubrics can be edited easily, and new rubrics can be developed using templates. This site is particularly useful for new teachers who may be unfamiliar with creating and using rubrics. If students know the criteria by which they will be judged, many students will actually work harder and perform better than if the criteria are unknown or vague. Providing students with a rubric before beginning an assignment allows them to know how their work will be evaluated.

To help students see the *learning* value of assessments, teachers must find ways to motivate students to review their assessments more than once. One way is for teachers to place a score on the paper, but not mark or explain what is incorrect. As homework, students must review the test and make corrections. Another way is to return a marked test and have students make corrections with an explanation of their incorrect answers and their correct answers. A third way is to give students the rubric and their essays and have them grade themselves or rewrite their answers. There are many creative ways teachers can engage students with, and learn from, their tests. Simply giving back a test and asking "any questions?" misses opportunities to teach and learn.

Scoring

There are at least five ways that teachers can increase the fairness with which tests are scored:

1. *Cover/block the student's name when grading student work.* This reduces bias and prevents possible personality issues from influencing the scoring of answers/work.

FIGURE 3.

Scoring Rubric for Essay Question

Four-Part Essay Prompt:

1. Write an operational question related to the experiment we did in class yesterday (in this case, "Cat's Meow"—food coloring in milk, add a drop of soap).

2. Develop a complete procedure to answer your question. Be sure your directions are clear and that you are answering the question you set out to answer!

3. Describe how your experiment is a controlled experiment. In other words, describe how your procedure was "fair."

4. Describe the data you would collect.

Criteria or Characteristics	High Quality	Medium Quality	Low Quality
How clear is the question investigated?	Question is easily understood. Question asks about relationship between two variables.	Question is easily understood. Question does not address relationship between two variables—e.g., may examine relationships between three variables.	Question is difficult to understand. Question does not address relationship between two variables—e.g., may examine relationships between three variables.
How well does the procedure address the question investigated?	The procedure clearly addresses the question investigated.	The procedure addresses the question investigated, but the reader has to assume one or two things to be certain.	The procedure does not seem to address the question investigated.
Was the test fair, or how well does the procedure control variables? [Many younger children will not be able to control variables yet.] Note: If the question investigated does not ask about the relationship between two variables, then this category cannot be assessed.	The student demonstrates that she or he consciously tries to control variables to make a fair test, uses multiple data points.	The student demonstrates that she or he consciously tried to control variables to make a fair test, but tried only two instances of the independent variable—e.g., tested only two temperatures.	The procedure does not control variables—i.e., not a fair test.

(continued)

How clearly was the procedure expressed?	Procedure is clear, succinct; reader can easily repeat procedure.	Procedure is moderately clear. Some details missing or muddled. Reader can repeat procedure with an educated guess about what to do at certain points.	Procedure is difficult to understand. May not correspond to what student did. Difficult to repeat, based on what was written.
Was the procedure repeated?	Procedure repeated enough times that student is confident of results.	Procedure repeated, but results conflicting. More trials required before student is confident of results.	Procedure was only conducted once.
What data are collected?	Student knows which data to collect and how to collect it. Extraneous data is not collected.	Student knows which data to collect but is not sure how to do so.	Student is not sure which data to collect or how to collect it. Is either vague about what data to collect (i.e., record all data and observations) or collects the wrong data.
Are accurate observations recorded?	*Student records accurate observations, including diagrams and written descriptions.*	*A written record is present, including diagrams or written descriptions, but not both.*	*No data were recorded by student, or student does not accurately describe observations. Data not intelligible to reader.*
Are observations appropriate to the question investigated?	*Student records observations appropriate to the question investigated.*	*A written record is present, but student does not describe observations appropriate to the question investigated.*	*No data are recorded by student, or irrelevant data recorded.*
Are the data displayed in a readable manner at a glance?	*Data are clear and easily understood at a glance. Data are organized using table, chart, graph, or similar communication method.*	*Data are understandable, but it takes some effort on part of reader to understand what student is saying.*	*No data are recorded by student, or data unintelligible.*
Is some sort of conclusion present?	*A conclusion is present.*	*A conclusion is present.*	*No conclusion is present.*

(continued)

Is the conclusion based on data from the experiment? Note: Obviously, this category cannot be assessed if no conclusion present; see previous category.	Student refers to and uses data to support a clear conclusion. Conclusion addresses original question.	Clear conclusion present, but connection between conclusion and data seems weak. Conclusion addresses original question.	Conclusion is present, but difficult for reader to understand how the conclusion is connected to data, or conclusion does not address original question.
Is there discussion of ways to continue or expand the investigation?	Report includes ways study could be continued or expanded.	Further discussion of investigateable questions, but questions only peripherally related to present study.	No discussion of ways study could be continued or expanded.

Note: The criteria and ratings in italics are used to evaluate how well the student actually conducted the experiment.

2. *Decide in advance what qualities are to be considered in judging student work.* For more traditional assessment strategies, an answer key can be used. For performance tasks or essays, the "ideal" answer, criteria for ideal responses, or a scoring rubric or scoring guide can be used.

3. *Grade or score all students' answers for one item/section before grading or scoring the next item.* This is especially important for constructed-response items for two reasons. First, a teacher may not be consistent on partial credit on all responses if that item is not scored at the same time. Scoring questions separately allows concentration on a single answer. Second, one must consider something called the "halo effect." Suppose a teacher grades a paper, and the student answers the question very well. When the teacher reads that same student's next response, there is a tendency to be more forgiving or lenient because the student performed so well on the first question. The impression of that student is formed and carried into the next question. The converse is also true. A student who answers a question poorly is often judged more harshly on the next item. If the student was unable to answer the first question, he or she must not have a clear idea for the next question (so goes the teacher's reasoning). By scoring one item at a time, teachers are less likely to carry over impressions of student work quality from one question to the next question.

4. *Reread test papers a second time, especially the first and last papers graded.* A limitation of essay questions is that grading is not always consistent. The papers graded at the beginning are often scrutinized more closely than papers graded later. Rereading answers allows a teacher to adjust scoring accordingly.

5. *Write comments on the papers.* This practice forces teachers to carefully read papers. It also provides proof to students that their work was carefully read

and makes it easier for teachers to remember aspects of the student answers; this is especially helpful when the answer has several parts.

When trying to decide which types of items to use, teachers should carefully consider their goals and the teaching strategies. The mode of assessment must align with *how* students were taught and be compatible with the goals of instruction. Tables 1 and 2 show goals for students, along with appropriate methods of assessing those goals.

One last issue to consider when reviewing tests is the allocation of scoring points in comparison to instruction. The percentage of points should be roughly equivalent to the amount of time spent teaching each concept. This can be difficult to achieve, as some topics are easily assessed and our tendency is to write more items on that topic. This can lead to the problem of omitting, or providing insufficient, items from other important topics taught in class.

TABLE 1.
Goals and Corresponding Assessment Strategies

| GOALS FOR STUDENTS | Multiple Choice | True/False | Short Answer | Matching | PERFORMANCE TASKS | | Observation Checklists | Peer Assessment | Graphic Organizers |
					Essays	Projects, Labs, or Reports			
Understand science content	X	X	X	X	X	X		X	X
Engage in cooperative work							X	X	
Display positive attitudes about science (like science)							X		
Apply knowledge to new situations or everyday life	X				X	X			
Be able generate and research questions					X	X			
Carry out scientific investigation					X	X			
Know about scientists and their inventions/discoveries	X	X	X	X	X	X			
Communicate scientific data					X	X			X
Be effective problem solvers					X	X			
Understand the nature of science	X	X	X	X	X	X			

TABLE 2.

Relative Merits of Selection-type Items and Supply-type Items

Characteristic	Selection-type Items (M-C, T/F, Matching)	Supply-type Items Short Answer	Essay
Measures factual information	Yes	Yes	Yes*
Measures understanding	Yes	No**	Yes
Measures synthesis	No**	No**	Yes
Easy to construct	No	Yes	Yes
Samples broadly	Yes	Yes	No
Eliminates bluffing	Yes	No	No
Eliminates writing skill	Yes	No	No
Eliminates blind guessing	No	Yes	Yes
Easy to score	Yes	No	No
Scoring is objective	Yes	No	No
Pinpoints learning errors	Yes	Yes	No
Encourages originality	No	No	Yes

* The essay test can measure knowledge of facts, but because of scoring and sampling problems, essays probably should not be used for this purpose.

**These items can be designed to measure limited aspects of these characteristics.

SUMMARY

Teachers need to document student growth and development. Formative assessment strategies help inform day-to-day instruction and let teachers determine students' levels of understanding. Summative assessment, the types of assessment items discussed in this chapter, are needed as part of the formal reporting of student work. Some types of assessment items (multiple choice, short answer, and essay) are useful for measuring specific student skills and knowledge. If students are conducting hands-on, inquiry investigations, there must also be assessment that documents those attributes and parameters. For some instructional goals, such as content knowledge and conceptual understanding, a "traditional" testing format may be more efficient than a performance task format. Teachers must carefully consider their goals, instructional strategies, and assessment approaches to locate a reasonable balance between the three areas.

APPENDIX

TRUE/FALSE ITEMS

Effective true/false items are very difficult to write because not all knowledge is measured well with the true/false format. Among the advantages of true/false items are that they make lower demands on reading abilities for students (than multiple-choice items), a large number of items can be answered in a typical testing period, and more true/false items can be asked than multiple-choice items. True/false and multiple-choice items are the least time consuming items for students to complete. Scoring by teachers is easier, considered more objective and reliable, and lends itself perfectly for learning outcomes when there are only two possible alternatives (categorical knowledge). The difficulty in writing true/false items beyond the categorical knowledge level is to write items that are free from ambiguity (see Figure 4 for evaluation suggestions). Limited diagnostic information can be gathered from incorrect responses by students. Typically, students who mark an item as false do not demonstrate that they know the answer, only that they recognize the statement as wrong. Scores are influenced by guessing more than with any other item type. Teachers can still learn a great deal from true/false items, however, because they can ask many questions in a given amount of time. Following are tips for writing true/false items.

1. Write true/false items in pairs. Be sure the content is parallel, with one statement being true and the other statement being false. By having a pair of statements, one true and one false, about the same content, teachers can select the item they want (true or false) so that they can construct a balanced test. It also allows the teacher to assess the content at two different times using different questions—using one version on a quiz and another on the test. Be sure the language is clear, brief, and stated simply.

2. Use absolute determiners wisely and sparingly (rarely does something *always* or *never* happen). Absolute determiners include *never, only, always, every,* and *all*; qualifying determiners include *generally, sometimes, usually, some, most, few,* and *often.*

3. Use more false items than true. Students are more likely to agree with something when they do not know an answer.

4. Use internal comparisons to help reduce ambiguity in statements (e.g., "Multiple-choice items allow for greater sampling of the content" would be aided by having an internal comparison: "Compared to essay questions, multiple-choice items allow for greater sampling of the content.").

FIGURE 4.

Evaluating True/False Items

Below are examples of problematic true/false items. The correct answer is noted and a critique of the item is on the right.

The flow of thermal energy (heat) between two objects always goes from hot to cold. (T)	A verbal clue is given—the flow of heat. This tells the test-wise student that heat is flowing and therefore must go from hot to cold. Specifying "hotter object to the colder object" instead of "hot to cold" would be better phrasing.
There are six planets in the solar system.	This is a great example of a problematic question because the question could logically be answered as either true or false. The question is vague—are there *only* six planets in the solar system or *at least* six? In which solar system—ours or some other one?
Multiple-choice items allow for more sampling of the content.	This is a great example of a problematic question because the question could logically be answered as either true or false. This statement needs an internal comparison—"more sampling than..." [i.e.,"more sampling than true/false items" (F); "more sampling than essay questions" (T)].
Alexander Graham Bell discovered the telephone. (F)	Devices are developed or invented but not discovered. The wording gives away the answer.
Newton invented gravity. (F)	People do not invent natural phenomena. They discover/determine how to harness their potential. The wording gives away the answer.

APPENDIX (continued)

MATCHING ITEMS

Matching items can be thought of as a series of efficiently presented multiple-choice items. Matching items are particularly useful for measuring factual information, such as dates of events, achievements of people, symbols for chemical elements, or names for parts of a diagram. They are effective for measuring associations, but less effective for measuring higher levels of understanding. Matching items provide a compact, efficient way to assess knowledge about a single topic, when reading and response time is relatively short. The items can be easily converted to/from multiple-choice items with a common set of foils. Scoring of matching items is relatively easy, objective, and reliable. Matching items work best for student knowledge outcomes based on association. Following are tips for writing matching items (see also Figure 5).

1. Keep the set of statements (stems) homogeneous.

2. Keep the list of stems and answers relatively short. This will also help with keeping the lists homogeneous. It is more effective to have two different sets of matching questions than to mix content types.

3. Use a heading for each column that describes its content.

4. Include more answers than questions, and/or allow students to use the same answer more than once. This decreases guessing and process-of-elimination responses and reduces the chance that a single incorrect answer forces multiple incorrect answers by students.

5. Ask students to choose answers from the list that has shorter statements. This minimizes time students spend on reading, since they will have to reread the list for each item.

6. Arrange the answers in a logical way (e.g., alphabetical, ascending/descending, chronological). This helps students who know the answer to locate it quickly.

7. Give clear directions. Reducing ambiguity always helps students perform better.

8. Arrange both lists on a single page. If the list is too long, consider making two sets of questions and answers.

FIGURE 5.
Evaluating Matching Items

Below are examples of matching items that are problematic. A critique of this set of items and answers follows the items.

1. What is the speed of light?	a. Sir Isaac Newton
2. Whom do we credit with first understanding gravitational theory?	b. 9.8 m/s^2
3. What is the relationship between wave speed and frequency?	c. $3.0 * 10^8 \text{ m/s}$
4. What is the acceleration due to gravity near the surface of the earth?	d. $v = f * \lambda$

There is a lack of consistency in question type and answers. There is only one of each type listed—one person, one speed, one formula. Students do not need to know the subject in order to do well on this matching test.

The bulk of the reading is on the left side. Students will re-read the list of answers multiple times as they search for the correct answer. Having the shorter answers on the right side means that they read the short text multiple times. This is the way you want it to be.

The items should be about similar content. This set of matching items contains concepts on gravity and light mixed together. Test-wise students will be able to select the correct answers without knowing the content.

LINKS TO THE NATIONAL SCIENCE EDUCATION STANDARDS

The National Science Education Standards (NRC 1996) call for certain levels of knowledge that all students must have. Teachers need to document student growth and development toward each Standard. Formative assessment strategies help inform day-to-day instruction and let teachers know where students are. Summative assessment, the types of assessment items discussed in this chapter, are needed as part of the formal reporting of student work.

No single form of assessment will adequately evaluate every one of our goals as teachers. The types of items discussed in this chapter (multiple choice, short answer, and essay) are useful for measuring some student skills and knowledge, but not all. If students are going to be conducting hands-on, inquiry investigations, there must be some assessment that "counts" those attributes. For some goals, content knowledge, and conceptual understanding, a traditional testing format may be more efficient than a performance task. Teachers need to consider carefully their goals, instructional strategies, and assessment approaches so that they find a balance between the three. As any one of them changes, so must the others.

RESOURCES

Ebel, R. L., and D. A. Frisbie. 1991. *Essentials of educational measurement*. Upper Saddle River, NJ: Prentice Hall.

Enger, S. K., and R. E. Yager. 2001. *Assessing student understanding in science*. Thousand Oaks, CA: Corwin..

Goodrich, H. 1997. Understanding rubrics. *Educational Leadership* 54 (4): 14–17.

Jones, M. G. 1994. Assessment potpourri. *Science and Children* 32 (2): 14–17.

Marzano, R. J., D. Pickering, and J. McTighe 1993. *Assessing student outcomes: Performance assessment using the dimensions of learning model*. Alexandria, VA: Association for Supervision and Curriculum Development.

Mertler, Craig A. 2001. Designing scoring rubrics for your classroom. *Practical Assessment, Research and Evaluation* 7 (25). Available online: *http://ericae.net/pare/getvn.asp?v=7&n=25*.

National Research Council (NRC). 2001. *Classroom assessment and the national science education standards*. Washington, DC: National Academy Press.

Thorndike, R. M., G. K. Cunningham, R. L. Thorndike, and E. P. Hagen. 1991. *Measurement and evaluation in psychology and education*. New York: Macmillan.

AUTHOR AFFILIATIONS

Laura Henriques is an associate professor and acting chair of science education at California State Long Beach.

Alan Colburn is the Master's Program coordinator and associate professor of science education at California State Long Beach.

William C. Ritz is the Head Start on Science director and professor emeritus of science education at California State Long Beach.

assessment in support of conceptual understanding and student motivation to learn science

Melissa DiGennaro King

INTRODUCTION

Education researchers have suggested that we must "fundamentally rethink the relationship between assessment and effective schooling" (Stiggins 1999, p. 1). Assessment traditions and conventional assessment practices in the United States have relied heavily on the belief that assessment for public accountability leads to academic improvement. The veracity of this notion is now being challenged (Mintzes, Wandersee, and Novak 2000; Stiggins 1999; Wiggins 1998).

The emphasis on external, high-stakes tests has dominated the assessment literature (Black and Wiliam 1998a). The growth of constructivism and the relatively stagnant performance of U.S. students in science (NAEP 2000) have fueled an important shift: a recent movement toward establishing a research-based rationale for improving classroom assessment (Mintzes, Wandersee, and Novak 2000; NRC 2001).

BACKGROUND

A review of studies conducted during and after 1988 indicates that more than 95% of students' experiences with educational assessment occur within the context of classroom instruction (Black and Wiliam 1998b; NRC 2001). Renewed interest in the interaction between teachers and students in various learning environments and a closer look at classroom-based assessment during the 1990s provides evidence that assessment is an integral aspect of practically every instructional event (NRC 2001). Recent calls for assessment that is continuous, cyclical, sensitive to individual differences, and reflective of deep understanding (NRC 1996; Stiggins 1999) have led to studies of alternative approaches that are embedded within the context of ongoing science instruction (Treagust et al. 2001).

As a follow-up from a comprehensive review of 250 articles that focused on educational assessment and student evaluation, Black and Wiliam (1998a) selected 40 controlled research studies for a detailed meta-analysis. Results of their research showed that formative, or embedded, assessment practices led to substantial gains in student learning. Among the studies reviewed, effect sizes for classroom assessment interventions ranged from 0.4 to 0.7, with low achievers benefiting the most from formative assessment (Black and Wiliam 1998a). These findings suggest that when effective formative assessment strategies are closely linked with regular instruction, student achievement may increase and the achievement gap may decrease. An important finding of this meta-analysis was that *the learner* should be closely involved with the process of formative assessment (Black and Wiliam 1998a, 1998b; NRC 2001). A learner should understand the instructional goals and be aware that his or her current knowledge and skills fall short of those goals. Then the learner can plan what actions to take to improve his or her knowledge and skills.

Other researchers have found that students tend to succeed academically when they believe they are capable of meeting learning challenges and are highly motivated (Stiggins 1999; Zimmerman and Schunk 2001). Studies have shown that self-efficacy is highly correlated with self-regulated learning, motivational beliefs, and student achievement (Zimmerman and Schunk 2001). Therefore, assessment strategies that build academic self-efficacy and motivation hold great promise as educational interventions (Black and Wiliam 1998b; Brookhart 1997).

Another team of researchers (Treagust et al. 2001) found that when teachers incorporate assessment events into daily instruction, they can extract and interpret information about what learners know and understand through careful observation, interactive dialogue, and varied expressions (e.g., writing and drawing) of scientific reasoning. These data provide teachers with valuable information about student progress and, in turn, help them as they make instructional changes to maximize the benefits of learning events (Brookhart 1997; Wiggins 1998). Research also suggests that students who receive feedback about their own performances and those who engage in self-assessment gain improved understanding of the steps they need to take to succeed academically (Black and Wiliam 1998a; NRC 2001; Zimmerman and Schunk 2001).

A major concern of today's educators and education researchers is, "How can we influence students' desire to learn and help learners take primary responsibility for academic success?" Students need positive experiences with learning in order to build self-confidence and to realize that learning is worth the effort expended (Stiggins 1999; Zimmerman and Schunk 2001). One researcher has suggested that "classroom assessment permeates all phases of teaching" (Brookhart 1997, p. 162). This point of view suggests that classroom assessment is a vehicle that links instruction with individual and group learning experiences. The classroom assessment environment serves as a sociocultural reality with deeply rooted effects on learners. The classroom climate, the teacher's expertise with assessment, and the teacher's beliefs about the teaching-learning process contribute to students' beliefs about assessment (Brookhart 1997). One research study indicates that assessment methods influence young learners' perceptions of themselves as students and the decisions they make to support their own learning (Thomas and Oldfather 1997). Over time, these factors may influence student achievement.

Formative assessment that takes place within the context of ongoing learning has been called a "naturalistic" approach because of its similarity with anthropological research (Thomas and Oldfather 1997). Teachers who make use of embedded assessment assume multiple roles at different times during instruction. They observe, interact, and monitor student performance while capturing data (both product and process) to measure student progress (Thomas and Oldfather 1997). Some research suggests that embedded assessment focused on meaningful understanding, rather than correct answers, encourages students to take risks, think creatively, engage in reflection, and discover their own strengths (Black and Wiliam 1998a; Treagust et al. 2001). However, accountability pressures during the past 15 years have narrowed the curriculum, and many of today's schools and classrooms emphasize "teaching to the test" rather than "teaching for understanding" (Schauble and Glaser 1996; Stiggins 1999; Wiggins 1998).

RESEARCH QUESTIONS, METHODOLOGY, AND ANALYSIS

The following questions guided the research study discussed in this chapter:

* What were children's perceptions about learning science?
* How did these elementary students feel about learning science? Why did they have those feelings?
* How did these students feel about being assessed in science? Why did they feel that way?
* What assessment strategies motivated these students to be interested in science?

The study took place in an urban elementary school in Northern Virginia and included three teachers and 65 students from three second-grade classrooms. The student population was predominantly white, middle-class (68%), but other ethnic groups were also represented (18% Hispanic, 8% Asian, 6% African American). The students included English language learners, special education students, and gifted students. Two of the participating teachers were highly experienced (each had taught for more than 20 years); the other teacher was relatively new to the profession (less than 3 years of teaching experience). The study spanned five weeks in the fall quarter of the school year. The research design emphasized an investigation of the *processes* for delivering and assessing elementary science. The researcher was a participant observer during science instruction in these classes.

She offered support to the second-grade teachers as they planned and implemented the district's science curriculum. She offered ideas for activities and assisted teachers in the delivery of hands-on experiences. She also developed a summative assessment instrument for use at the end of the target science unit. After observing student performance on that assessment instrument, she conducted individual interviews with six second-grade students selected from the three second-grade classrooms.

Qualitative methods (Maxwell 1996) were employed, with an intentional focus on building research relationships with participants in the study (teachers and students). The goal of the research was to gain greater understanding about young children's perspectives on learning and assessment in science. Initially, the researcher concentrated on establishing rapport with students. During the course of the study, she was on site several times a week, observing and participating in science instruction. Moving beyond "stranger status" was critically important for success with the individual interviews, which occurred at the end of the instructional unit.

The researcher also developed relationships with teachers by providing support for science instruction. The three teachers in the study had minimal science backgrounds and did not convey a high degree of confidence in their abilities to teach science. The researcher offered demonstration lessons, materials, and resources. This collegial support ensured that everyone gained something valuable. Over time, the researcher's presence in the classrooms became more comfortable and natural, and classroom visits were not perceived as intrusions. One participating teacher remarked to the researcher, "When you come into my classroom you really help, and I don't even have to tell you what to do." The careful negotiation of research relationships *in situ* and the purposeful establishment of reciprocal gains for all participants continued throughout the study.

The main sources of data for this study were detailed field notes, classroom observation summaries, and verbatim transcripts of student interviews. The initial qualitative analysis involved reviewing the transcripts and written descriptions from each individual interview. This allowed the researcher to discover commonalities and search for critical differences in children's responses. Children delivered multiple messages through their words, actions, facial expressions, body language, use of pauses, and spontaneous questions. With young children, what they *do not* say can be as important as what *do* they say. After reviewing these data, the researcher developed "emic" categories (Maxwell 1996) to represent the conceptual structures that seemed to characterize the thoughts and feelings of these young learners. Using concept mapping techniques, the researcher created a visual diagram for these categories and their relationships (see Figure 1). A qualitative coding scheme was then developed for analysis of the text from the student interviews. The concept map displays the main ideas of children's expressions and shows how they are connected: (a) positive attitude toward science, (b) motivation to learn, (c) assessment, and (d) conceptual understanding.

FIGURE 1.

Concept Map Connecting Positive Attitude Toward Science ("Like Science"), Motivation to Learn, Assessment, and Conceptual Understanding

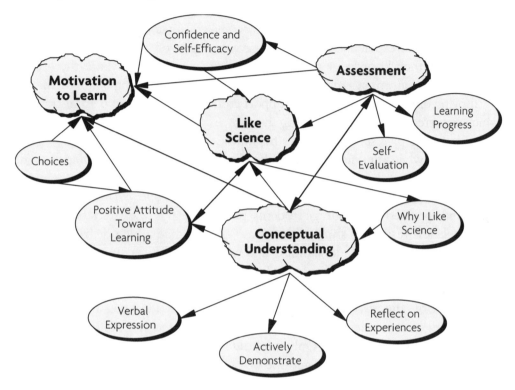

The researcher created a matrix chart with numerical data to indicate the number of responses in each coded category for each interviewee. This process helped to determine emerging patterns as well as any unique or unusual features. Examination of these data led to the construction of a coherent story to interpret and explain children's voices.

FINDINGS AND IMPLICATIONS FOR CLASSROOM USE

The second-grade students developed strong positive attitudes toward learning science. They demonstrated abundant enthusiasm when they exclaimed, "I like science a lot" and "I think science is fun." Several children repeated comments such as, "It's interesting" and "It's fun!" during the interviews, just to make sure that the researcher got the message. The children enjoyed the active involvement of hands-on science, as expressed in their own words: "You can do a lot of experiments and stuff!" and "I like science because it's a lot to do." The second graders liked the element of surprise in science, saying excitedly, "Wow! I never knew that!" and "In science, you never know what's going to happen." The children seemed to view science as an intriguing mystery that unfolded through unique experiences with "a ton of stuff."

After completion of the unit, these young learners demonstrated basic understanding of magnetism and provided rudimentary explanations of magnetic force. Furthermore, they relished the opportunity to share their knowledge with the researcher as they used appropriate vocabulary, analytical reasoning, and logical justifications for their explanations. Students gave examples of concrete experiences to demonstrate their knowledge and understanding—for example, about the orderly arrangement of atoms in a magnet. Some students needed to manipulate real materials (moving objects and magnets) to support recall. These hands-on demonstrations, which occurred spontaneously during the course of some interviews, functioned as self-assessments or tests of children's hypotheses, conclusions, and interpretations. For example, while interviewees manipulated materials and magnets during dialogue with the researcher, they created opportunities to check on whether or not their own explanations were verifiable. These continuing interactions allowed for further exploration and appeared to encourage students' positive feelings toward science.

Interview responses showed that these children had a high degree of confidence in their levels of understanding and were highly efficacious about science. These students believed that they had the ability to be successful with the science assessment tasks and activities. Among the comments reflecting these beliefs were "Most of [the test] was pretty easy because we had learned a lot about it," "I'm doing good [in science] 'cause it's not so hard," and "I feel like I might get everything right [on the test]." Only one student revealed any anxiety about being tested when she said, "I might get one or two wrong." Several children actually expressed enthusiasm about being tested because it gave them an opportunity to show what they had learned.

The students expressed a preference for open-ended test questions. They explained that open-ended probes allowed them to demonstrate understanding with fewer constraints. One student said, "After I draw a picture, it's easier to share what I know." Another commented, "You could think of a little solution and then write it down." Students' comments indicated that having choices and multiple options for assessment responses were also appealing to them. One child said, "Writing and drawing was better than just drawing because then you could tell what I was drawing [even] better." Another said, "I like how they give you a choice." The seven-year-olds seemed to be aware of their own strengths and weaknesses in communicating conceptual understanding. Some students described specific test-taking strategies that they had applied, such as, "Because I tried them all out in one sentence, and then it makes sense." Interaction between the researcher and the learners revealed that the summative assessment for the unit on magnetism was a positive experience for these students.

EXAMPLES OF CLASSROOM APPLICATIONS

On the test, students were asked whether or not magnetic force goes through air, and they were directed to draw a picture to show their explanations. This process required that children reflect on a concrete experience and relate this event to a logical explanation of the phenomenon observed. Earlier, the children watched how a paper clip, tied to a string and secured to a tabletop with tape, was suspended in mid-air when a wand magnet was held nearby. Although the drawings were rudimentary, the students explained what was happening when given opportunities to talk about their illustrations. Some degree of abstract reasoning may be available to children when it is linked with direct, coherent experiences. However, in order for children to share what they learned, they needed access to multiple means of expression. For example, limitations in children's drawing abilities meant that pictures were not sufficient for capturing their understandings of the principles of magnetic force. They also needed verbal or written communication to express themselves more clearly.

In another example, students were presented with a drawing and told that it was incorrect. In the drawing, paper clips were stuck to the center of the horseshoe magnet, rather than being attracted to the poles (the ends of the horseshoe magnet). The students were directed to make a scientifically accurate drawing and to write one to two sentences explaining why they made specific changes to their drawings. The children were delighted to fix something that was wrong, and they smiled as they proceeded. For some students, however, vocabulary became a stumbling block during writing (e.g., some could not remember that the ends of a horseshoe magnet were called North and South Poles). When given opportunities to express ideas verbally, most children referred to "stronger force" at the magnetic poles, supplementing oral language with distinct gestures to demonstrate knowledge of concepts, such as strength. These responses illustrated the complex nature of young children's scientific sense making. Assessment that offers students multiple options for self-expression provides greater possibilities for students to communicate what they understand. These demonstrations showed that developmental constraints may be operating when young children's levels of conceptual understanding do not match their capacity to use written or graphical conventions (e.g., writing or drawing).

On the summative assessment, students were also asked to show the arrangement of atoms inside an iron nail that had been magnetized. Instructional activities included lining students up in the classroom to show the orderly (versus disorderly) pattern of atomic particles that influence magnetic force. A rhythmic song with simple lyrics reinforced this concept. Children placed small Post-it notes on two large pieces of paper (shaped like nails) taped on the wall to show how atoms lined up in magnetized objects. This placement of Post-it notes in organized rows or in a disorganized fashion was a simple way to determine which children understood the target concept. Going one step further on the summative assessment, second graders were asked to create graphic representations of the idea. Although this specific concept was beyond the requirements of the second-grade curriculum, the students succeeded on this task on the end-of-unit test and seemed to understand how it related to magnetic force.

In summary, these findings suggest that young children may grasp abstract scientific concepts when they have had firsthand experiences that clarify and personalize meaning in concrete ways. Assessment strategies may serve as bridges that support the transition from concrete experiences to other symbolic representations and higher levels of thinking.

IMPLICATIONS FOR SCIENCE ASSESSMENT

This study of assessment in an elementary classroom context suggests that young children enjoy science learning. Their positive attitudes may have developed from opportunities for active exploration and the intellectual stimulation that resulted from new and unexpected discoveries in science. These pleasurable experiences boosted academic motivation and students' beliefs in their abilities to "do" science. These beliefs and feelings helped these young learners approach testing situations with little or no anxiety. Successful formative and summative assessment experiences may have contributed to their continued enthusiasm for science. Additional studies are needed to determine the validity of these interpretations.

Findings from this study suggest that assessments that include elements of choice (e.g., "choose which question you want to answer") and multiple options for demonstrating understanding (e.g., drawing, writing, verbalizing) are powerful ways to engage young children's thinking. Practitioners should listen carefully to children's "voices" as they consider ways to measure content knowledge. In this study, children's actions and comments showed that they viewed assessment as a natural part of learning. These second graders did not feel threatened by tests, perhaps because they believed in their own abilities to be successful when engaged in science experiences. Establishing a classroom culture that includes ongoing assessment as a routine and natural part of instruction may have beneficial outcomes for learning (Brookhart 1997; Wiggins 1998). Further research is needed.

LINKS TO THE NATIONAL SCIENCE EDUCATION STANDARDS

The National Science Education Standards (NRC 1996) state that "teachers are in the best position to put assessment data to powerful use" (p. 87). The research study discussed here is linked with Teaching Standard C of the Standards. Assessment data should be matched with decisions and actions. Assessment Standard C includes an emphasis on providing students with ample opportunity to demonstrate their achievements.

REFERENCES

Black, P., and D. Wiliam. 1998a. Assessment and classroom learning. *Assessment in Education* 5 (1): 7–74.

Black, P., and D. Wiliam. 1998b. Inside the black box: Raising standards through classroom assessment. *Phi Delta Kappan* 80 (2): 139–148.

Brookhart, S. 1997. A theoretical framework for the role of classroom assessment in motivating student effort and achievement. *Applied Measurement in Education* 10 (2): 161–180.

Maxwell, J. 1996. *Qualitative research design: An interactive approach.* Thousand Oaks, CA: Sage.

Mintzes, J., J. Wandersee, and J. Novak. 2000. *Assessing science understanding.* San Diego, CA: Academic Press.

National assessment of educational progress (NAEP). 2000. Washington, DC: National Center for Education Statistics. Available at *http://nces.gov/nationsreportcard.*

National Research Council (NRC). 1996. *National science education standards.* Washington, DC: National Academy Press.

National Research Council (NRC). 2001. *Classroom assessment and the national science education standards.* Washington, DC: National Academy Press.

Schauble, L., and R. Glaser. 1996. *Innovations in learning: New environments for education.* Mahwah, NJ: Lawrence Erlbaum.

Stiggins, R. 1999. Assessment, student confidence, and school success. *Phi Delta Kappan* 81 (3): 191–198.

Thomas, S., and P. Oldfather. 1997. Intrinsic motivation, literacy, and assessment practices: That's my grade, that's me. *Educational Psychologist* 32 (2): 107–123.

Treagust, D., R. Jacobowitz, J. Gallagher, and J. Parker. 2001. Using assessment as a guide in teaching for understanding: A case study of a middle school science class learning about sound. *Science Education* 85: 137–157.

Wiggins, G. 1998. *Educative assessment.* San Francisco, CA: Jossey-Bass.

Zimmerman, B. J., and D. H. Schunk. 2001. *Self-regulated learning and academic achievement: Theoretical perspectives.* Mahwah, NJ: Lawrence Erlbaum.

RESOURCES
Print

Brown, J. H., and R. J. Shavelson 1996. *Assessing hands-on science.* Thousand Oaks, CA: Corwin.

Burke, K. 1999. *How to assess authentic learning.* Arlington Heights, IL: Skylight.

National Science Foundation (NSF). 1999. *Inquiry: Thoughts, views, and strategies for the K–5 classroom.* Arlington, VA: NSF, Directorate for Education and Human Resources.

Stiggins, R. 1997. *Student-centered classroom assessment.* Upper Saddle River, NJ: Prentice Hall.

Web

www.enc.org/topics/assessment/classroom

http://ericae.net/pare

www.cse.ucla.edu/cresst

www.pals.sri.com

AUTHOR AFFILIATIONS
Melissa King is a teaching specialist at K12, Inc., an educational publishing company that develops online and print materials for grades K–12. She has a doctoral degree in science education from George Mason University (GMU). She has been an elementary and a middle school teacher in Arlington County, Virginia, and has taught graduate-level methods courses at GMU and the University of Virginia. King also serves as a distance-learning instructor for the graduate program at Kaplan University.

adaptive inquiry as the silver bullet: reconciling local curriculum, instruction, and assessment procedures with state-mandated testing in science

Stephen J. Farenga, Beverly A. Joyce, and Daniel Ness

INTRODUCTION

During the past decade, changes in curriculum, delivery of instruction, and assessment have altered the landscape of science and mathematics education. Such patterns of change have recurred throughout history (Broudy and Palmer 1965). The purpose of a particular methodological approach to teaching is dependent on what a particular society values at any given time. Moreover, national and state standards reflect the current education climate.

BACKGROUND

Current changes in curriculum, instruction, and assessment that have been endorsed by national and state organizations (e.g., see Table 1) demonstrate the shift in science and mathematics education from an authoritarian, direct approach to a collaborative, inquiry-based approach. Some observers suggest that these changes are due to a confluence of societal demands: international comparisons on achievement (such as the Third International

Mathematics and Science Study [TIMSS] [Johnson et al. 2005]), the needs of the U.S industrial complex, and national security concerns (NRC 1996; NCES 1998).

TABLE 1.
National Science Education Standards Changing Emphases in Teaching

The National Science Education Standards envision change throughout the system. The teaching standards encompass the following changes in emphases:

Less Emphasis On	More Emphasis On
Treating all students alike and responding to the group as a whole	Understanding and responding to individual students' interest, strengths, experiences, and needs
Rigidly following curriculum	Selecting and adapting curriculum
Focusing on student acquisition of information	Focusing on student understanding and use of scientific knowledge, ideas, and inquiry processes
Presenting scientific knowledge through lecture, text, and demonstration	Guiding students in active and extended scientific inquiry
Asking for recitation of acquired knowledge	Providing opportunities for scientific discussion and debate among students
Testing students for factual information at the end of the unit or chapter	Continuously assessing student understanding
Maintaining responsibility and authority	Sharing responsibility for learning with students
Supporting competition	Supporting a classroom community with cooperation, shared responsibility, and respect
Working alone	Working with other teachers to enhance the science program

Source: National Research Council (NRC). 1996. National science education standards. Washington, DC: National Academy Press, p. 52.

Methods of pedagogy have been under scrutiny for over two millennia, as evidenced by the educational and epistemological traditions emanating from the seminal writings of Socrates, Plato, Comenius, Pestalozzi, Rousseau, Herbart, and Piaget. One can view themes of pedagogical instruction in terms of a pendulum metaphor. At one point in history, direct instruction is the *sine qua non* method; at another point, discovery teaching is seen as imperative in the classroom. An argument for the need of methodological fluidity can be made by comparing the New York public schools of the 1930s with the schools of today. Methods that then served as a successful solution for educating immigrant populations may not be of value now.

Moreover, methods have sometimes been applied to all students in varied environments. This myopic view of a one-size-fits-all program usually leads to short-term, limited success and eventual abandonment. That was the fate of the "alphabet soup" science programs of the 1960s and 1970s (e.g., Science Curriculum Improvement Study; Science: A Process Approach; and the Elementary Science Study).

The pressure created when students do not achieve success causes teachers to continually search for new forms of instructional penicillin. Before teachers inquire about the most effective method to teach a particular topic, however, they must first address the purpose of what is to be learned. The purpose should then direct the form of instruction, the curriculum, and the method of assessment to be outlined by local, state, or national agencies.

More than 55 years ago, Ralph Tyler was one of the first educational theorists to unify instruction, curriculum, and assessment. According to Tyler's theory, teachers and education agencies should ask the following questions prior to creating curricular instructional strategies and methods of assessment:

* What educational purposes should the school seek to attain?
* What educational experiences can be provided that are likely to attain these purposes?
* How can these educational experiences be effectively organized?
* How can we determine whether these purposes are being attained?

Tyler's questions have been useful for developing frameworks for evaluating the purpose of instruction. To illustrate this, we developed a scenario (see below) regarding instructional practices, curriculum experiences, and assessment methods through clinical observations and interviews with middle school science teachers. The scenario was a composite of best practices in seventh-grade science as demonstrated by "Ms. Wallace." The Standards for Professional Development for Teachers of Science (NRC 1996) maintain that a career-long professional development plan must include self-reflective, inquiry-based activities. We suggest that teachers reflect on their own teaching practices while analyzing what was done by Ms. Wallace.

Ms. Wallace's Seventh-Grade Classroom

Ms. Wallace taught middle school science in New York State for 17 years. She worked collaboratively with her colleagues to establish an inquiry-based science program. Ms. Wallace often engaged her students in real-life problem solving, as evidenced by her unit on properties of matter. She had students carry out activities to familiarize them with scale and structure of matter. She stressed the properties of solids, liquids, and gases while also addressing the questions about the properties of matter that the students wanted to investigate. In the unit's closing lesson, students worked in their usual groups of four to investigate a discrepant event, using cornstarch, water, and food coloring. This activity gave students the opportunity to experience a substance that appeared to defy what had been learned about Newton's law regarding the viscosity of a fluid. Upon further investigation, students realized that the viscosity of the substance could be changed by applying force or stress. This observation brought students to question Newton's law, which states that the viscosity of a fluid can only be changed by varying the fluid's temperature. Ms. Wallace then had students design a test and identify as many variables as possible for the investigation.

Ms. Wallace's objectives for the unit were to have students develop skills in the following areas: observing and recording data, using scientific reasoning to understand relationships among variables, analyzing their own learning and developing a plan for future learning, working in teams to conduct research, designing assessment techniques to identify the properties of the non-Newtonian fluids, and questioning their scientific understanding.

Ms. Wallace used embedded assessment techniques during instruction. By questioning individual students and students in small groups, she assessed students' initial understanding and abilities, monitored progress, and collected achievement data. Her assessment procedures included formative measures such as anecdotal records, questioning, checklists, quizzes, and peer- and self-evaluation. Her primary instructional approach was to have students work in groups, teams, or pairs to solve problems. It was evident that she established a supportive classroom community since students were engaged in collaborative learning and shared responsibility, as envisioned by the National Science Education Standards.

FINDINGS AND DISCUSSION

A TEACHER'S CONUNDRUM: INSTRUCTION AND ASSESSMENT AT VARIANCE WITH STATE SCIENCE TESTS

Although Ms. Wallace's science lessons embodied the essence of the National Science Education Standards (see Table 2), further analysis is needed to identify achievement in science understanding by students. Ms. Wallace's display of pedagogical-content knowledge and assessment methods seemed to work well in terms of her daily classroom procedures. Students worked in collaborative learning groups, participated on research teams, and discussed outcomes with lab partners. Her methods met the recommended National Science Education Standards. However, her methods did not match the required skills set necessary for the New York State-mandated eighth-grade high-stakes science test, called the Intermediate-Level Science Examination (University of the State of New York 2000). That test requires students to work independently on both its written and manipulative

TABLE 2.

Links to the National Science Education Standards in Ms. Wallace's Classes

Teaching Standard	What Ms. Wallace Did
A: Teachers of science plan an inquiry-based science program for their students. (NRC*, p. 30)	Ms. Wallace collaborated with her colleagues to plan an inquiry-based science program.
B: Teachers of science guide and facilitate learning. (NRC, p. 32)	Ms. Wallace gave students multiple opportunities to learn the objective of the lesson. They obtained knowledge of the properties of matter.
C: Teachers of science engage in ongoing assessment of their teaching and of student learning. (NRC, p. 37)	Ms. Wallace asked questions. Ms. Wallace kept anecdotal records of students' work.
D: Teachers of science design and manage learning environments that provide students with the time, space, and resources needed for learning science. (NRC, p. 43)	Students worked in pairs, teams, and groups. Materials were supplied to conduct an inquiry-based investigation.
E: Teachers of science develop communities of science learners that reflect the intellectual rigor of scientific inquiry and the attitudes and social values conducive to science learning. (NRC, p. 45)	Students worked in pairs, teams, and groups. They were provided with problems that challenge conventional thought, thus providing fodder for further questions and inquiry. Through discrepant events, students were taught to question their scientific understanding.
F: Teachers of science actively participate in the ongoing planning and development of the school science program. (NRC, p. 51)	Ms. Wallace actively engaged in planning the science program in her school.
Assessment Standard	What Mrs. Wallace Did
A: Assessments must be consistent with the decisions they are designed to inform. (NRC, p. 78)	Ms. Wallace used multiple forms of assessment to gather a complete understanding of the students' abilities (e.g., anecdotal records and discourse with students).
B: Achievement and opportunity to learn science must be addressed. (NRC, p. 79)	Students planned and conducted their own science investigations.
C: The technical quality of the data collected is well matched to the decisions and actions taken on the basis of their interpretation. (NRC, p. 83)	Ms. Wallace tried to ensure that students could clearly identify and accurately describe the properties of matter prior to continuing with the next topic.

* NRC = National Research Council. 1996. National science education standards. Washington, DC: National Academy Press.

skills sections. It does not allow students to problem solve collaboratively, participate on research teams, or discuss outcomes. Clearly, the New York State summative assessment experience was diametrically opposed to that of Ms. Wallace's formative assessment and instructional practices.

As a classroom teacher, Ms. Wallace plays a key role in the alignment of curriculum, instruction, and assessment. Cognizant of the significant impact of New York State's standardized assessment program and the local school district's assigned curriculum, Ms. Wallace outlined and executed a flexible process of instruction that responded to changing student needs as the class progressed from direct instruction to open-ended inquiry. She used the data from formative assessments to establish levels of prior knowledge, shape student behavior, correct misinterpreted information, and modify her teaching approaches. It was critical that her instruction fed compatibly into curriculum and assessment to form a "best fit" model (Figure 1), thereby acknowledging the interdependent status of curriculum, instruction, and assessment. This approach allowed Ms. Wallace to effectively address the necessary course content without limiting her instruction to "teaching to the test."

Valid assessment—both formative and summative—must be aligned with the content, process, and product of the classroom experience. Formative assessment techniques, such as anecdotal notes, checklists, quizzes, and self- and peer-evaluations, are initiated by the teacher and provide samples of behavior that indicate a student's mastery level of content and procedures. Effective use of formative assessment as a key element in the instruction process ensures that students will be familiar with—not surprised by—the nature of the science problems encountered on the summative assessment (i.e., the Intermediate-Level Science Examination). Since the scope, content, and difficulty level of that test have been externally defined, developed, endorsed, and prescribed by New York State, the challenge to the classroom teacher is to develop compatible methods of instruction to prepare students for the high-stakes examination.

A basic premise of assessment validity is that the instrument mirrors the alignment of content, process, and product experienced in the classroom. If an element of the method of assessment is the antithesis of the method of instruction, then the validity of assessment results is questionable. An attempt to follow opposing directives (in this case, the New York State Intermediate-Level Science Examination) places Ms. Wallace and her colleagues face-to-face with a conundrum.

ADAPTIVE INQUIRY AS A WAY TO RESOLVE THE CONUNDRUM

Policy makers, along with teachers, must begin with a clear definition of inquiry-based instruction. Currently, a broad definition of inquiry includes general processes of exploring phenomena that lead to questions, discoveries, hypotheses, and tests to support or refute original thoughts. It is obvious that this definition of inquiry can apply to a variety of pedagogical approaches in science teaching. At the classroom level, teachers must engage in action research to find evidence to support their actions in the face of the instructional-assessment conundrum. Teachers determine the extent to which the effectiveness of instruction depends on the kinds of students taught. In addition, they must determine if the method of instruction is consistent with how students will be assessed.

Many science educators think that the National Science Education Standards insist that the sole emphasis of instruction should be on collaborative, student

FIGURE 1.

A Model for Aligning Curriculum, Instruction, and Assessment

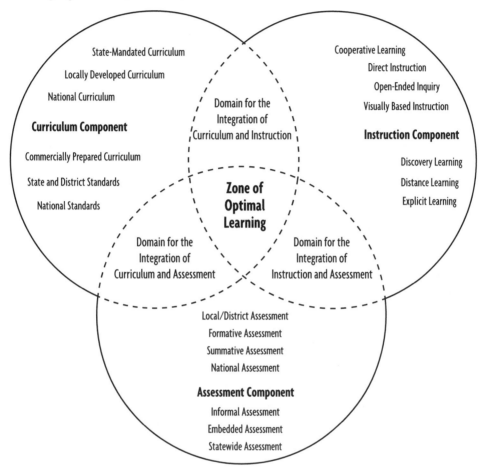

Figure adapted from Farenga, S. J., B. A. Joyce, and D. Ness. 2002. Reaching the Zone of Optimal Learning: The alignment of curriculum, instruction, and assessment. In Learning science and the science of learning, *ed. R. W. Bybee. Arlington, VA: NSTA Press.*

directed, inquiry-based investigations. This is understandable since the literature is replete with readings promoting inquiry-based learning (Layman, Ochoa, and Heikkinen 1996; NRC 1999; Saul and Reardon 1996; Whitin and Whitin 1997). However, careful examination of the Standards shows that "teachers of science must decide when and for what purpose to use whole-class instruction, small-group collaboration, and individual work" (p. 32). Even though individual work is mentioned, it is evident that the spirit of the Standards fosters the creation of communities of science learners in which teachers facilitate inquiry-based activities. Furthermore, current research on teaching does not support individual learning and direct instruction as effective means of learning science.

One method to allow a teacher to tailor a teaching strategy to the cognitive levels of each student is the method of instruction called adaptive inquiry (Farenga and Joyce 2002; Farenga, Joyce, and Ness 2002). We developed adaptive inquiry in light of the need to address the demands of science assessment in New York State. This new form of assessment required teachers to reexamine curriculum, instructional techniques, and measures of achievement. Evidence suggested that effective teachers need to possess a broad pedagogical-content repertoire that varies from direct instruction to open-ended inquiry in order to enhance student achievement on summative assessments. Bransford, Brown, and Cocking (2000) identify the importance of having a variety of pedagogical teaching strategies. Teachers who are fluid in the use of these strategies recognize the need to align curriculum, instruction, and assessment in order to reach optimal learning conditions (Farenga, Joyce, and Ness 2002).

Research in science and mathematics education is generally split when considering most effective teaching strategies. On the one hand, open-ended inquiry has been found within a number of studies to be an effective teaching strategy (Pressley and McCormick 1995; Rakow 1986; Shulman and Keislar 1966; White and Frederickson 1998), while other studies find direct instruction to be more effective than other strategies (Becker and Gersten 1982; Klahr and Nigam 2004; Meyer 1984; Meyer, Gersten, and Gutkin 1984). However, when each strategy is used exclusively, it may not fulfill its instructional potential (Farenga, Joyce, and Ness 2002; Klahr and Nigam 2004; Ness 2002). Teachers need to assess prior knowledge and design the delivery of instruction to match the cognitive level of the student. The teacher must then assess student responses to instruction and move students from concrete understandings to more abstract conceptual development. Subsequently, the teacher must consistently process the information obtained from instruction, students' outcomes, and prior knowledge; this continuous feedback and interpretation of information is the basis of adaptive inquiry.

Adaptive inquiry is the product of the synergistic relationship between what a student brings to the classroom, the teacher's ability to shape a lesson in response to the needs of the student, and the method of final assessment. The lesson is based on students' prior knowledge and what they need and want to know. Flexibility in the design, assessment, and communication of results is paramount to promote the concept of science as inquiry-at-work.

In support of adaptive inquiry, recent studies in cognitive science suggest that methods of instruction need to vary based on students' experiential and knowledge levels (Landauer and Dumais 1996; Larkin, et al. 1980; Miller and Gildea 1987; Pressley and McCormick 1995; Shulman and Keislar 1966; Tuovinen and Sweller 1999). Students who lack adequate factual or domain knowledge in a specific content area may not be best served by a discovery approach to teaching. Tuovinen and Sweller (1999, p. 340) suggest that "exploration practice clearly caused a much larger cognitive load and led to poor[er] learning" for students with limited pre-existing knowledge. An optimal method of instruction for these students was to acquire factual knowledge through examples, to develop and strengthen basic cognitive schemata for the particular content area learned. Acquiring this knowledge base fostered the development of cognitive schemata that handled new content efficiently, while lessening the demand on the student's working memory. When inquiry-based instruction supplanted direct instruction as an initial method of teaching, students with limited factual knowledge could find the exploration associated with discovery or inquiry to be an ineffective primary means of learning. [In this chapter, we use the terms *inquiry* and *discovery learning*. In reviewing the research, we found that a number of studies did not define specific approaches for each method. There was a pedagogical difference between inquiry and discovery learning (the difference lay in who identified the problem under investigation). In authentic inquiry, more of the decisions regarding investigation were made by the students.]

Direct instruction necessitates that the curriculum be carefully sequenced throughout the instructional process. This form of instruction appears to facilitate learning for students with limited experience in the subject being taught. In this case, the teacher mediates and controls the learning activities and processes. Conversely, the locus of control with the inquiry model is situated with the student. Using inquiry as a form of teaching and learning requires students to begin with a question, design an investigation, develop a hypothesis, collect and use data to answer the original question, determine whether the original question requires modification, and communicate results. Students with limited knowledge in a specific domain may find these tasks difficult to carry out. Therefore, students who participate in summative tests (e. g., New York's Intermediate-Level Science Examination) that are different in content and process from classroom practice may suffer experiential deficits in science content or process skills, leading to low levels of achievement.

Research from the cognitive sciences may provide powerful evidence that one particular method of instruction may not be the best means of learning for all students. Standardized tests (such as the Intermediate-Level Science Examination), which are used to compare students against state performance standards, may not match what is taught in all classrooms. The scope of what is learned in inquiry science is so extensive that it is difficult to construct an assessment instrument that parallels what teachers may be doing in their classrooms. Indeed, a broader range of instructional practices that match a comprehensive list of assessment procedures is needed. In our opinion, methods such as adaptive inquiry can be used to address the varying cognitive and developmental strengths of individual learners. Further, the use of adaptive inquiry supports the premise that instructional methods should combine formative assessment practices of the teacher with summative assessment practices from external education agencies.

Acknowledgment

The authors wish to thank Mark J. Diercks, a middle school science teacher, for his identification of problems related to opposing directives from two external sources in developing classroom practices.

Links to the National Science Education Standards

Ms. Wallace's instructional approach indicates that she embraces the National Science Education Standards (NSES) (NRC 1996) for understanding the nature of the learner, the nature of instruction, and the method of assessment (see Table 2). This was evidenced by her involvement in the development of science programs and planning of science lessons (Teaching Standard A and F). The content of the lessons were aligned with the National Science Physical Science Content Standards for grades 5–8 (Content Standard B). The discussion of properties and changes in properties of matter was taught using an inquiry-based approach (Content Standard A). Ms. Wallace guided and facilitated student learning (Teaching Standard B) while she assessed what was learned and how the content of the lesson was delivered (Teaching Standard C). Moreover, she fostered the growth of learning communities (Teaching Standard D) by encouraging students to work together in collaborative groups (Teaching Standard E). Ms. Wallace coordinated formative assessment techniques with their intended purpose (Assessment Standard A) while measuring students' achievement and their opportunities to learn (Assessment Standard B). In addition, her assessment tasks were embedded in student-based activities, where she collected appropriate data to support student outcomes (Assessment Standard C).

References

Becker, W. C., and R. Gersten. 1982. A follow up of Follow Through: The later effects of the direct instruction model on children in fifth and sixth grades. *American Educational Research Journal* 19: 75–92.

Bransford, J. D., A. L. Brown, and R. R. Cocking. 2000. *How people learn: Brain, mind, experience, and school.* Washington, DC: National Academy Press.

Broudy, H. S., and J. R. Palmer. 1965. *Exemplars of teaching method.* Chicago: Rand McNally.

Farenga, S. J., and B. A. Joyce. 2002. Teaching youngsters science in a culturally diverse urban classroom. In *Commitment to excellence: Transforming teaching and teacher education in inner-city and urban settings,* eds. A. C. Diver-Stamnes and L. A. Catelli, 149–170. Cresskill, NJ: Hampton Press.

Farenga, S. J., B. A. Joyce, and T. Dowling. 2002. Adaptive inquiry: Building launch compressor rockets. *Science Scope* 25 (4): 34–39.

Farenga, S. J., B. A. Joyce, and D. Ness. 2002. Reaching the zone of optimal learning: Aligning curriculum, instruction, and assessment. In *Learning science and the science of learning,* ed. R. Bybee, 51–64. Arlington, VA: NSTA Press.

Johnson, E., J. Cohen, W. H. Chen, T. Jiang, and Y. Zhang. 2005. *2000 NAEP–1999 TIMSS linking report.* Washington, DC: National Center for Education Statistics.

Klahr, D., and M. Nigam. 2004. The equivalence of learning paths in early science instruction: Effects of direct instruction and discovery learning. *Psychological Science* 15(10): 661–667.

Landauer, T., and S. Dumais. 1996. How come you know so much? From practical problems to new memory theory. In *Basic and applied memory research,* eds. D. J. Hermann, C. McEvoy, C. Hertzog, P. Hertel, and M. K. Johnson. Mahwah, NJ: Lawrence Erlbaum.

Larkin, J., J. McDermott, D. P. Simon, and H. A. Simon. 1980. Expert and novice performance in solving physics problems. *Science* 208: 1335–1342.

Layman, J. W., G. Ochoa, and H. Heikkinen. 1996. *Inquiry and learning: Realizing science standards in the classroom*. New York: College Entrance Examination Board.

Meyer, L. 1984. Long-term academic effects of Direct Instruction Follow Through. *Elementary School Journal* 4: 380–394.

Meyer, L., R. Gersten, and J. Gutkin. 1984. Direct Instruction: A Project Follow Through success story. *Elementary School Journal* 2: 241–252.

Miller, G. A., and P. Gildea. 1987. How children learn words. *Scientific American* (Sept.): 94–99.

National Center for Education Statistics (NCES). 1998. *Pursuing excellence*. Washington, DC: U. S. Department of Education.

National Research Council (NRC). 1996. *National science education standards*. Washington, DC: National Academy Press.

National Research Council (NRC). 1999. *Inquiry and the national science education standards: A guide for teaching and learning*. Washington, DC: National Academy Press.

Ness, D. 2002. Helping teachers recognize and connect the culturally bound nature of young children's mathematical intuitions to in-school mathematics concepts. In *Commitment to excellence: Transforming teaching and teacher education in inner-city and urban settings*, eds. A. C. Diver-Stamnes and L. A. Catelli, 171–189. Cresskill, NJ: Hampton Press.

Pressley, M., and C. McCormick. 1995. *Cognition, teaching, and assessment*. New York: Harper Collins.

Rakow, S. J. 1986. *Teaching science as inquiry*. Bloomington, IN: Phi Delta Kappa.

Saul, W., and J. Reardon, eds. 1996. *Beyond the science kit: Inquiry in action*. Portsmouth, NH: Heinemann.

Shulman, L. S., and E. R. Keislar. 1966. *Learning by discovery: A critical appraisal*. Chicago: Rand McNally.

Tuovinen, J. E., and J. Sweller. 1999. A comparison of cognitive load associated with discovery learning and worked examples. *Journal of Educational Psychology* 91(2): 334–341.

Tyler, R. 1949. *Basic principles of curriculum and instruction*. Chicago: University of Chicago Press.

University of the State of New York. 2000. *Intermediate-level science examination*. Albany, NY: University of the State of New York.

White, B. Y., and J. R. Frederickson. 1998. Inquiry, modeling, and metacognition: Making science accessible to all students. *Cognition and Instruction* 16 (1): 3–117.

Whitin, P., and D. J. Whitin. 1997. *Inquiry at the window*. Portsmouth, NH: Heinemann.

AUTHOR AFFILIATIONS

Stephen J. Farenga is an associate professor of human development and learning, **Beverly A. Joyce** is an associate professor of human development and learning, and **Daniel Ness** is an assistant professor of human development and learning—all in the School of Education at Dowling College.

science standards influence classroom assessment practices

Kathy J. McWaters and Ronald G. Good

INTRODUCTION

Do the science standards influence classroom assessment practices? If so, to what extent? If not, why not? What are the classroom assessment practices of teachers? Who else has asked these questions? What did they find out? What contribution will the answers make to science education? This research project represented the beginning of a journey to seek answers to these questions and share what we have learned.

BACKGROUND

The research project that we describe in this chapter had two purposes: (1) to identify the classroom assessment practices of middle school (grades 5–8) science teachers and (2) to explore the relationship between these practices and those recommended in national and state science standards. To address these issues, in 1999–2000 we reviewed relevant and current literature. With few exceptions (e.g., Cizek, Fitzgerald, and Rachor 1995; Shepardson and Adams 1996), most research studies designed to understand science teachers' classroom assessment practices in the United States were published before 1989, predating the publication of current national and state science reform documents. Consequently, such studies were not focused on the influence of national and state standards on teachers' classroom assessment practices.

Prior research findings on classroom assessment practices indicate that teachers and measurement specialists do not share the same knowledge base. Many teachers view assessment as being summative—separate and apart from teaching. Furthermore, according to researchers, student acceptance, the nature and quality of the information obtained from the test, external constraints placed on teachers, and the ways in which teachers intend to use test data may be as important, or more important in some cases, than the technical quality of the test in influencing classroom assessment practices (Morais and Miranda 1996; Wilson 1989). The results of a few studies (Kamen 1996; Ruiz-Primo and Shavelson 1996; Borko, Flory, and Cumbo 1993) indicate that nontraditional forms of assessment offer a useful way for teachers to integrate teaching and assessment, and that some teachers are successful at using them.

The research base about the influence of national standards on assessment is still in the early stages. Some might argue that it is too soon to study the influence of the standards on teachers' classroom practices. However, as the authors of *Science for All Americans* (AAAS 1989) suggested, if people are still talking about and working toward science reform in ten years we will have been successful.

RESEARCH QUESTIONS, METHODS, AND ANALYSIS

The following research questions were addressed as part of this study:

* What are the classroom assessment practices (CAPs) of middle school science teachers?
* To what extent do national or state science standards influence CAPs?

Sub-questions included the following: Are teachers providing students with the opportunity to learn the material in national or state content standards? Do teachers' views on science teaching and learning support the vision of science learning set forth in science reform documents?

The study was conducted in two phases, using a mix of quantitative and qualitative research methods described by Tashakkori and Teddlie (1998). In Phase 1, the researcher mailed a questionnaire (the Classroom Assessment Practices Questionnaire, or CAPQ) to 450 middle school (grades 5–8) science teachers in 17 parishes in Louisiana to obtain information about their classroom assessment practices. Forty-four percent (197) of the questionnaires were returned. In Phase 2, eight middle school teachers in eight departmentalized classrooms, two classes at each grade, participated in a study to investigate the classroom assessment practices of middle school science teachers.

Quantitative data were analyzed using the teacher as the unit of analysis (Tashakkori and Teddlie 1998). Descriptive statistics were used to analyze the responses to the CAPQs. The qualitative data were analyzed before the researcher analyzed the questionnaire data. Doing the analysis in this order prevented the researcher from being influenced by the survey results.

Classroom observations by the researcher (54 hours), face-to-face teacher interviews (18 interviews), and analyses of classroom assessment documents (eight documents) generated data that were analyzed using cross-case analysis. (This meant grouping together answers from different people and analyzing different perspectives on the central issues as described by Patton 1990). Interviewing and observing teachers across multiple grades (fifth, sixth, seventh, and eighth) provided the researcher with a rich opportunity to study classroom assessment practices. By combining observations and interviewing techniques, the researcher checked with teachers to verify the intent of their assessment behaviors.

Teacher-made tests were analyzed according to a modified version of an assessment analysis procedure (see Appendix) developed by members of Project 2061 (G. Kulm, personal communication, June 27, 1997). This procedure had four steps: identification of the standards and benchmarks addressed by the test questions, determination of content alignment, examination of the test format, and production of a summary.

RESULTS

AVAILABILITY AND USE OF SCIENCE REFORM DOCUMENTS

Teachers who had copies of the national, state, and/or regional science reform documents claimed to have used them (see Table 1). About three-fourths of respondents used at least one science reform document. However, nearly one-fourth of the teachers surveyed reported not having and/or not using any of the reform documents. These data make the case for asking the school librarian to set aside space for a professional library that could contain current sets of reform documents.

TABLE 1.

Availability and Use of Science Reform Documents by Middle School Science Teachers*

	Availability of Documents		Use of Documents	
	% of Teachers	Number of Teachers	% of Teachers	Number of Teachers
No documents	23	45	—	—
One or more documents	77	152	76	150

*Total number of teachers = 197 (450 questionnaires mailed to teachers; 197 returned)

Regarding teachers' use of national and state documents, 149 teachers reported using a national document and 140 teachers reported using a state document (see Table 2). Obviously, these categories were not mutually exclusive. Questionnaire data indicated that teachers who used science standards tended to have and use more than one reform document.

During interviews, the researcher asked teachers to identify (from a display of books) which science reform documents they used. The teachers most often

TABLE 2.
Teachers' Use of Specific Science Reform Documents*

Document	Use		Do Not Use	
	% of Teachers	Number of Teachers	% of Teachers	Number of Teachers
National				
Science for All Americans	2	4	98	193
National Science Education Standards	30	60	70	137
Benchmarks for Science Literacy	43	85	57	112
State				
Louisiana Science Framework	56	110	44	87
Teachers' Guide to Statewide Science Assessment, Grades 4, 8, and 11	15	30	85	167

* Total number of teachers = 197

indicated the *Louisiana Science Framework* (state document), *ScienceWorks* (regional document), and their local school district's concept grids. Teachers did not recognize the actual state assessment reform document, *Teachers' Guide to Science Assessment, Grades 4, 8, and 10* (Louisiana Department of Education 1998). Both the state and the district had sent copies of the guide to the school principals. Additionally, the district superintendent had sent a directive to the principals requiring them to copy and distribute the assessment document to the teachers. As the interviews progressed, the researcher discovered that only part of the complete document had been distributed to each teacher. Moreover, only the eighth-grade teachers saw any value in the document. The fifth-, sixth-, and seventh-grade science teachers were mainly concerned with the content and format of the Iowa Test of Basic Skills (ITBS) taken by their students at the end of the school year.

TEACHERS' USE OF STANDARDS IN ASSESSMENT CONSTRUCTION AND SELECTION

Teachers were asked on the CAPQ to "briefly provide an example of how you use the science standards when constructing or selecting assessments." Eight distinct categories emerged from the data (Table 3). Thirty-five percent (69) of the teachers did not respond to the question, so it was not known how or whether they used the science standards when constructing or selecting assessments.

Additionally, 5% (10) of teachers indicated that they did not use the standards when constructing assessments. Twenty-five percent (49) of the remaining

TABLE 3.
Teachers' Responses Regarding Use of Standards in Assessment Construction and Selection*

Response Categories	% of Teachers	Number of Teachers
1. Focusing on the format of the test		
Tests are developed based on standards.	9	17
Standards are used to justify tests.	4	8
Response does not indicate whether the test or the standard "came first."	3	6
2. Making a general statement linking standards to the assessment instrument		
Standards are some type of general guide.	4	7
Teachers use standards because they are pressured to do so.	13	26
3. Indicating that standards are used to write lesson plans and objectives.	6	11
4. Not addressing the issue of standards	5	10
5. Not relating to assessment	7	13
6. Giving specific examples of techniques used to correlate standards with assessment	10	20
7. Indicating that teachers do not use standards in developing assessments	5	10
8. No response	35	69

Total number of teachers = 197

responses were statements that did not relate standards to assessment selection or construction, or indicated that teachers were paying "lip service" to the standards because they were pressured to do so.

Thirty-five percent (69) of the teachers indicated that they used the standards in the selection and construction of classroom assessments. The most direct use of the standards was to select content, format, and cognitive level for test items. A more circumspect approach by teachers was using the standards to write lesson plans and objectives, and then plan assessments based on these teacher-constructed objectives. The problem created by doing this was that the "big ideas" of the standards were reduced by the specific, measurable objectives that many teachers were required to write. The resulting assessment became one that focused on factual data, without the rich context envisioned in the standards, and reinforced students' conceptions of science as a body of knowledge to be memorized.

ALIGNMENT BETWEEN STANDARDS AND TEACHER-MADE TESTS

All of the eight teachers who were observed believed that they planned their instruction and their assessments to align with the national, state, and/or regional science content standards. The manner in which each teacher translated the standards into teaching and assessment strategies was different. Even the most traditional of teachers (straight rows and busy, quiet students, working individually) had the parish concept grid in their lesson plan books, and they assured the researcher (during an interview) that everything taught was based on the standards.

Three of the eight teachers emphasized with students the understanding of particular science concepts and engaged students in partial inquiry methods during instruction; nevertheless, they tested students on facts and recall. This finding was in agreement with Canady and Hotchkiss (1989), who found that teachers emphasized higher-order-thinking skills during instruction yet tested students on facts and recall. The teachers in our study clearly taught according to standards and benchmarks and used recommended teaching strategies. However, they quite literally stopped teaching science after the unit test was constructed and began helping students memorize questions and answers. We have termed this behavior "test rehearsal." This practice was observed in both fifth-grade classrooms and one sixth-grade classroom.

It is important to note that each teacher would have been given "high marks" in terms of teaching appropriate science content using effective strategies. It was only when the researcher analyzed the alignment among the standards, benchmarks, lesson plans, and teacher-made tests that the disconnect between the quality of teaching and the quality of testing was recognized. The roots of the disconnect seemed to lie in the lack of teachers' assessment literacy. An assessment analysis matrix (Table 4) shows the alignment between standards and teacher-made tests.

Three other kinds of assessment difficulties were identified—two related to the lack of effective teaching strategies and the third related to the lack of time spent in science class. The first assessment issue occurred in one seventh-grade class, in which the teacher wrote outlines on the chalkboard and then had the students copy them. As the teacher went though the outlines with the students, she made every effort to make the class "interesting and fun," but she relied on the use of rote memory techniques (such as mnemonics) to help her students memorize information.

The second assessment issue was the direct result of ineffective teaching methods. In this eighth-grade classroom, students sat in rows and were not allowed to talk. The only voice heard in the room was that of the 30-year veteran teacher. Her comments concentrated on the correction of student behavior. The only other time the teacher spoke was to give directions and correct students' written questions. Twice a week students checked their work by reading aloud the answers to questions they had written during earlier classes. Students were expected to "work" and get "it" (correct answers) in class or at home. Worksheets were used daily. The unit test was provided by the textbook publisher. Test questions had been answered by students on earlier worksheets. In order to "pass the

TABLE 4.
Assessment Analysis Matrix

Criteria	Teacher-Made Test		
	Teacher A (Grade 8)	Teacher B (Grade 7)	Teacher A (Grade 6)
Content Alignment to the Standards			
Substance	X	P	P
Sophistication	X	X	P
Coverage	P	P	P
Extraneous content	O	O	O
Format Analysis in Terms of the Standards			
Fit of item format	X	X	NE
Important ideas	X	P	NE
Context and fairness	X	X	NE
Usefulness	X	P	NE

Note: X = good alignment; P = partial alignment; O = not present; NE = not evaluated

test," students memorized information from their worksheets. This teacher used an outdated textbook, not the textbook adopted by the school district.

The third assessment issue revolved around the amount of time students spent in science class. In one sixth grade, the science class was either frequently canceled because of the teacher's professional leave or because the school activity schedule required students to be out of class.

Alignment between the tests used by teachers and the National Science Education Standards and Benchmarks for Science Literacy was examined using a draft version of Project 2061's Assessment Analysis (see Appendix). The content alignment included examining the link between the test and the teachers' learning goals and the match between the teachers' learning goals and the benchmarks listed in national, state, and regional documents. The purpose of the format analysis was to estimate how well the test addressed the standards and benchmarks based on the criteria of: the fit of individual item format, important ideas, context and fairness, and usefulness. Findings from this study suggested that some problems with teachers' classroom assessment practices related to ineffective teaching practices.

CONCLUSIONS

In our opinion, middle school science teachers' assessment practices are influenced by state science content standards and benchmarks, and to a lesser degree by the National Science Education Standards and the Benchmarks for Science Literacy. The results from this study show that the degree of influence varies from teacher to teacher. If a standards-based instruction/assessment program is important, we need to understand what happens to the "big ideas" of science as national content standards and benchmarks are converted into state content standards and benchmarks, and then translated into teachers' lesson plan objectives.

Curriculum development has been used successfully as a vehicle for staff development. Research is needed to determine if the development of assessment systems can also be effectively used as a vehicle for staff development. Two teachers who participated in this study indicated an interest in continuing to work with the authors to develop teacher research projects in their classrooms. They wanted to develop a set of multiple assessment activities and a record-keeping system that reflected where their students were in the journey toward achieving the seven content standards of the National Science Education Standards, without overwhelming themselves or their students in mounds of paperwork.

In addition to reading science reform documents, how can teachers develop a greater understanding of assessment issues and their relationship to science learning? One way may be to use Project 2061's Assessment Analysis (Appendix) as a basis for beginning assessment conversations among teachers and between teachers and researchers. The essential characteristics of exemplary assessment practices are included in the National Science Education Standards (Figure 1) and are offered as guides for the development of assessment tasks and practices.

Both preservice and inservice teachers need time to discuss what it means to align their lesson objectives with the standards and benchmarks and with their own testing practices. To be effective, teachers need the opportunity to study the National Science Education Standards and Benchmarks for Science Literacy and work with colleagues in their districts and schools to make local curriculum decisions about which benchmarks will be taught and how attainment of the standards will be measured. To improve communication among the teachers, principal, and district staff, teachers may want to consider establishing a school curriculum team to work with the principal and district curriculum staff to align the teachers' assessments with the standards.

Finally, and perhaps most importantly, we all need to continue looking for ways to bridge the gap between research and the classroom by engaging in collaborative studies between classroom teachers and university researchers. Teachers continue to identify and try out what has come to be known as "best practice" for effective science teaching, learning, and assessing in their classrooms.

FIGURE 1.

The National Science Education Assessment Standards

Standard A	Assessments must be consistent with the decisions they are designed to inform.	* Assessments are deliberately designed. * Assessments have explicitly stated purposes. * The relationship between the decisions and the data is clear. * Assessment procedures are internally consistent. (p. 78)
Standard B	Achievement and opportunity to learn science must be assessed.	* Achievement data collected focus on the science content that is most important for students to learn. * Opportunity-to-learn data collected focus on the most powerful indicators. * Equal attention must be given to the assessment of opportunity to learn and to the assessment of student achievement. (p. 79)
Standard C	The technical quality of the data collected is well matched to the decisions and actions taken on the basis of their interpretation.	* The feature that is claimed to be measured is actually measured. * Assessment tasks are authentic. * An individual student's performance is similar on two or more tasks that claim to measure the same aspect of student achievement. * Students have adequate opportunity to demonstrate their achievements. * Assessment tasks and methods of presenting them provide data that are sufficiently stable to lead to the same decisions if used at different times. (p. 83)
Standard D	Assessment practices must be fair.	* Assessment tasks must be reviewed for the use of stereotypes, for assumptions that reflect the perspectives or experiences of a particular group, for language that might be offensive to a particular group, and for other features that might distract students from the intended task. * Large-scale assessments must use statistical techniques to identify potential bias among subgroups. * Assessment tasks must be appropriately modified to accommodate the needs of students with physical disabilities, learning disabilities, or limited English proficiency. * Assessment tasks must be set in a variety of contexts, be engaging to students with different interests and experiences, and must not assume the perspective or experience of a particular gender, racial, or ethnic group. (p. 85)
Standard E	The inferences made from assessments about student achievement and opportunity to learn must be sound.	* When making inferences from assessment data about student achievement and opportunity to learn science, explicit reference needs to be made to the assumptions on which the inferences are based. (p. 86)

Source: National Research Council (NRC). 1996. National science education standards. Washington, DC: National Academy Press.

Appendix
Project 2061 Assessment Analysis (Draft Document 1997)

Preliminary Evaluation

The analysis begins with an item or test that appears promising—the content doesn't appear too far outside the scope of science literacy and it contains at least some open-response opportunities. The task is to identify a collection of specific learning goals (benchmarks or standards) that appear to be central to the item or test.

First, reviewers search fairly quickly through the item or test (both questions and scoring guides or rubrics) to make a preliminary list of all the specific learning goals that would seem likely to be targeted. The assessment is then examined more carefully to determine whether each learning goal is actually seen—e.g., particular items, questions, and performances. Then, based on the analysis, benchmarks and standards are ranked from high to low to give a rough picture of how well they are addressed in the item or test.

Content Alignment

This analysis is a more rigorous examination of the link between the assessment and the selected learning goals. It proceeds in one or two stages, depending on the length of the assessment.

If the analysis is of a single item, possibly with a few sub-questions, the content analysis consists of an intense investigation of the match between one or two key benchmarks and the item. The first step involves giving precise attention to both ends of the match in order to clarify the specific ideas and skills that are included in the benchmarks of interest and to identify which of those ideas and skills receive significant attention in the assessment under study. Answers are sought to such questions as:

> *Substance*—Does the item address the specific substance of a benchmark or is there only a general "topic" correspondence? Does the item address specific content knowledge or does it mainly require only general knowledge? Does the item require additional specialized knowledge not included in the benchmark?

> *Sophistication*—Does the item reflect the level of sophistication of the benchmark or is it more appropriate for assessing benchmarks at an earlier or later grade level?

> *Part/Whole*—Does the item address all elements of a benchmark, or only some parts? If the latter, what is the consequence?

If the analysis is of an entire assessment, a subtest that consists of several items, or a long multi-step item that addresses several benchmarks, the first step is followed by a second one, which surveys the assessment as a whole. The purpose of surveying the test as a whole is to estimate the degree of overlap between its content and the learning goals of interest. This analysis addresses questions such as these:

> *Coverage*—What set of benchmarks for a given topic and grade level are addressed by the test? Which, if any, are not treated? Are the missing benchmarks essential?

> *Extraneous Content*—Does the test contain content—knowledge and skills—not required for reaching science literacy learning goals? If so, in what proportion?

Format Analysis

The purpose of the format analysis is to estimate how well the item addresses the central benchmarks from the perspective of what is known about student learning and effective assessment. The criteria for making such judgments are derived from research on assessment and from experience in classrooms and testing. Four criteria have been identified to serve as a basis for the format analysis. Stated as questions, these are:

> *Fit of Item Format*—Does the question (open-response, multiple choice, etc.) fit the type of knowledge assessed (e.g., skill, knowledge, application)? Does the type of question provide an opportunity to determine whether or not a student has actually met the relevant learning goal?

> *Important Ideas*—Are the scientific or mathematical ideas important, and are they given more attention in the item than reading skills, or recall of technical terms, vocabulary, or symbols? Does the item address central ideas rather than isolated pieces of information? Does the item require students to apply big ideas explaining phenomenon, or in making inferences or deductions?

> *Context and Fairness*—Does the item use a familiar, realistic, or meaningful setting that is relevant to the

targeted benchmark? Does the item help students see that the question or problem is important to address? Does the item use a context or situation that is familiar to all backgrounds and to both genders? Is the language clear and readable?

Usefulness—Does the item or scoring rubric provide information that would be useful to the student, teacher, or others in finding out about progress toward the learning goal or how to improve future instruction?

Profile

Having completed both a content analysis and a format analysis, the final step in the process is to prepare a profile that summarizes the main features of the subject material. Such a profile will include the conclusions reached with regard to (1) the treatment of key benchmarks, and (2) the overall character of the material. Even though the profile includes judgments regarding how well individual learning goals are treated, pointing to various strengths and weaknesses in the subject material, it does not conclude with a final, overall rating.

LINKS TO THE NATIONAL SCIENCE EDUCATION STANDARDS

"Begun in 1992, the *National Science Education Standards* (NSES) (NRC 1996) is part of the U.S. government's approach to education reform, an approach that involves setting national goals and the standards for meeting them" (DeBoer 2000, p. 590). The NSES offer content standards as learning goals for students just as the benchmarks of the American Association for the Advancement of Science do. The study reported on in this chapter looked for links between middle school science teachers' classroom assessment practices and national and state science content standards and benchmarks (learning goals for students). While specific links between assessment practices and learning goals are noted in the results and conclusions sections, it might be helpful to note that the analyses of the teachers' classroom assessment documents were partially aligned with NSES assessment standards: Standard B—Achievement and opportunity to learn science must be assessed; Standard C—The technical quality of the data collected is well matched to the decisions and actions taken on the basis of their interpretation; and, Standard D—Assessment practices must be fair.

REFERENCES

American Association for the Advancement of Science/Project 2061 (AAAS). 1989. *Science for all Americans.* New York: Oxford University Press.

American Association for the Advancement of Science/Project 2061 (AAAS). 1993. *Benchmarks for science literacy.* New York: Oxford University Press.

American Association for the Advancement of Science/Project 2061. 1997. October. Blueprints on-line. Available at *www.project2061.org/publications/bfr/online/blpintro.htm.*

Borko, H., M. Flory, and K. Cumbo. 1993. *Teachers' ideas and practices about assessment and instruction* (CSE Tech. Rep. No. 366). Los Angeles: University of California, National Center for Research on Evaluation, Standards, and Student Testing.

Canady, R. L., and P. R. Hotchkiss. 1989. It's a good score! Just a bad grade. *Phi Delta Kappan* 71 (1): 68–71.

Cizek, G. J., S. M. Fitzgerald, and R. E. Rachor. 1995. Teachers' assessment practices: Preparation, isolation, and the kitchen sink. *Educational Assessment* 3 (2): 159–179.

DeBoer, G. 2000. Scientific literacy: Another look at its historical and contemporary meanings and its relationship to science education reform. *Journal of Research in Science Teaching* 37 (6): 582–601.

Kamen, M. 1996. A teacher's implementation of authentic assessment in an elementary science classroom. *Journal of Research in Science Teaching* 33: 859–877.

Louisiana Department of Education. 1997. *Louisiana science framework: State standards for curriculum development*. Baton Rouge: Louisiana Department of Education.

Louisiana Department of Education. 1998. *Teachers' guide to statewide assessment in science, grades 4, 5, and 11*. Baton Rouge: Louisiana Department of Education.

McMorris, R. F., and R. A. Boothroyd. 1993. Tests that teachers build: An analysis of classroom tests in science and mathematics. *Applied Measurement in Education* 6 (4): 321–342.

Morais, A. M., and C. Miranda. 1996. Understanding teachers' evaluation criteria: A condition for success in science class. *Journal of Research in Science Teaching* 33: 601–624.

National Research Council (NRC). 1996. *National science education standards*. Washington, DC: National Academy Press.

Patton, M. Q. 1990. *Qualitative evaluation and research methods*. Newbury Park, CA: Sage.

Ruiz-Primo, M. A., and R. J. Shavelson. 1996. Rhetoric and reality in science performance assessments: An update. *Journal of Research in Science Teaching* 33: 1045–1063.

Shepardson, D. P., and P. E. Adams. 1996. Perspectives on assessment in science: Voices from the field. Paper presented at the meeting of the National Association for Research in Science Teaching, St. Louis, MO (March).

Tashakkori, A., and C. Teddlie. 1998. *Mixed methodology: Combining qualitative and quantitative approaches*. Newbury Park, CA: Sage.

Wilson, R.J. 1989. Evaluating student achievement in an Ontario high school. *The Alberta Journal of Educational Research* 35: 134–144.

Resources

Must-Read Books for Science Teachers

Doran, R., F. Chan, P. Tamir, and C. Lenhardt. 2002. *Science educator's guide to laboratory assessment*. Arlington, VA: NSTA Press.

Mintzes, J. M., J. H. Wandersee, and J. D. Novak, eds. 1997. *Teaching science for understanding: A human constructivist view*. San Diego: Academic Press.

Mintzes, J. M., J. H. Wandersee, and J. D. Novak, eds. 2000. *Assessing science understanding: A human constructivist view*. San Diego: Academic Press.

Oescher, J., and P. Kirby. 1990. Assessing teacher-made tests in secondary math and science classrooms. Paper presented at the meeting of the National Council on Measurement in Education, Boston, MA (April).

Spradley, J. P. 1980. *Participant observation*. Fort Worth: Harcourt Brace Jovanovich.

Internet Sites

www.project2061.org

This site is sponsored by Project 2061 (very user friendly).

www.nap.edu/catalog/9847.html

Classroom Assessment and the National Science Education Standards. 2001. J. M. Atkin, P. Black, and J. Coffey, eds. Committee on Classroom Assessment and the *National Science Education Standards*, Center for Education, National Research Council.

www.educ.sfu.ca/narstsite

The National Association for Research in Science Teaching (NARST) is a worldwide organization of professionals committed to the improvement of science teaching and learning through research. Since its inception in 1928, NARST has promoted research in science education and the communication of knowledge generated by the research.

www.nsta.org

The National Science Teachers Association homepage.

Journal Articles

Craven III, J. A., and T. Hogan. 2001. Assessing student participation in the classroom. *Science Scope* (Sept.): 36–40.

Farenga, S., and B. Joyce. 2000. Preparing for parents' questions. *Science Scope* (March): 12–14.

AUTHOR AFFILIATIONS

Kathy J. McWaters taught science for more than 20 years to middle school and high school students and is currently a supervisor for curriculum and special projects with the St. John the Baptist Public School System in Louisiana.

Ronald G. Good retired as a professor from Louisiana State University in Baton Rouge. He served as editor of the *Journal of Research in Science Teaching* from 1990 to 1993.

usable assessments for teaching science content and inquiry standards

Christopher J. Harris, Katherine L. McNeill, David J. Lizotte, Ronald W. Marx, and Joseph Krajcik

INTRODUCTION

Assessment is a critical means for determining the extent to which students achieve learning goals. The closer that student assessment is aligned with curriculum and classroom practice, the more likely it is that assessment data will provide an accurate picture of learning (Ruiz-Primo et al. 2002). Large-scale, standardized assessments are often separated from real classroom learning (Shepard 2000). Because these assessments are not aligned with classroom curriculum and instruction, they are less likely to be sensitive to changes in student learning. In contrast, assessments that align closely with curriculum and learning goals may be more immediately usable by teachers and researchers, especially for getting feedback about whether students are achieving goals and for adjusting curriculum and instruction accordingly.

In our work, we shift the focus of assessment toward the classroom, where the teaching and learning occurs. Assessments should be usable; they should be practical and informative for teachers and researchers. The effectiveness of an assessment depends, in large part, on how well it aligns with curriculum and instruction to reinforce common learning goals (Pellegrino, Chudowsky, and Glaser 2001). Our interest is in creating a strong alignment between assessment, curriculum, and learning goals through a process we call learning-goals-driven design.

BACKGROUND

Science education reform efforts have called for students to develop scientific practices and skills through inquiry (American Association for the Advancement of Science [AAAS] 1993; National Research Council [NRC] 1996). The National Science Education Standards strongly emphasize developing students' inquiry abilities, through which students should learn more than vocabulary definitions and content knowledge (NRC 1996). Content knowledge is certainly valuable; however, in reform-oriented, inquiry-based science classrooms, students are expected to participate in scientific practices and apply scientific ideas through describing and explaining phenomena, carrying out experiments and investigations, and collecting and analyzing data (Krajcik, Czerniak, and Berger 2003). In these science classrooms, knowing science does not simply mean that students remember facts; rather, students employ higher cognitive processes as they apply their knowledge in inquiry activities. While many forms of instruction appear to have the same effect on student learning when the only measure is factual recall, differences in learning outcomes become clear when multiple measures, including higher cognitive ones, are used (Bransford, Brown, and Cocking 2000).

Using only multiple-choice assessment does not access the deep, rich understandings called for in reform. Assessment should measure what students learn, and this learning should parallel the curriculum and the standards (Pellegrino, Chudowsky, and Glaser 2001). Although alignment can be difficult to achieve, classroom assessments aligned with learning goals and the curriculum hold promise for providing informative and useful feedback for teachers and researchers.

Both science teachers and researchers share the common goal of all designers—to develop courses of action aimed at modifying existing situations into preferred ones (Simon 1996), in this case increased student learning. Our research group has been developing middle school science instructional materials based on national standards using a learning-goals-driven design model (Reiseret al. 2003), based on Wiggins and McTighe's (1998) backward design model. Often, science teachers and researchers, with a general notion of what they want students to learn, will start with favorite activities or a textbook when they are designing curriculum. Instead, we began with our desired learning outcomes. Our commitment is that learning-goals-driven design can create better alignment between curriculum and assessment. We also think that this alignment can help us uncover more fine-tuned changes in student learning.

USING LEARNING-GOALS-DRIVEN DESIGN TO CREATE AN INSTRUCTIONAL UNIT
We used learning-goals-driven design to create a seventh-grade, inquiry-oriented chemistry unit called "How Can I Make New Stuff From Old Stuff?" (referred to as "New Stuff From Old Stuff"). The design process involved seven steps:

1. Identifying and clarifying national standards
2. Developing learning performances to meet standards
3. Creating base and specific assessment rubrics and accompanying assessment tasks

4. Designing learning tasks
5. Contextualizing content in relevant and everyday phenomena
6. Producing an instructional sequence including both student and teacher materials
7. Pilot testing materials and receiving feedback from external reviewers, including staff members of Project 2061 of the American Association for the Advancement of Science and science content reviewers.

Although these steps are listed linearly, in practice they were iterative. That is, the later steps of the design cycle, such as the assessments, informed previous steps, such as the learning performances (see Figure 1).

As the first step in our design process, we began with national standards to identify key middle school chemistry ideas. We focused on national content standards that addressed substances and properties (NRC 1996, p. 154) and chemical reactions (AAAS 1990, p. 47). Once we identified the relevant content standards,

FIGURE 1.

Learning-Goals-Driven Design Process

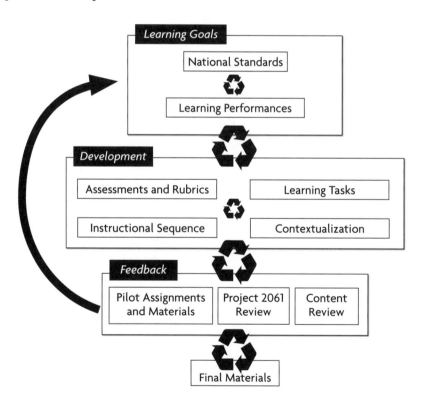

we unpacked these relatively succinct statements to clarify the science behind them (see Table 1).

For our second step we developed a range of learning performances that specified how students should be able to use the scientific ideas outlined in the standards. Building on Perkins's notion of understanding performances (Perkins 1998), we moved from the standards (a description of the scientific ideas) to per-

TABLE 1.

From National Science Standards* to Learning Performance

National Science Standard	Clarifying the Standard	Learning Performance (LP)
A substance has characteristic properties, such as density, a boiling point, and solubility, all of which are independent of the amount of the sample. (NRC 1996, p. 154)	A substance is made of one material throughout. Substances have distinct properties that can be used to distinguish and separate one substance from another. Properties such as density, melting point, and solubility describe the unique characteristics of substances. The properties of a substance do not change regardless of the amount of the substance. Density is the ratio of mass per unit volume. Melting point is the temperature at which a solid changes to a liquid. Solubility is the capacity of a solid to dissolve in a liquid.	LP1—Define a *substance* as something that is made of the same type of material throughout and a *mixture* as something that contains more than one type of material.
		LP2—Define a *property* as a unique characteristic that helps identify a substance and distinguish one substance from another. A property does not depend on the amount of the substance. A property of a substance is always the same.
		LP3—Identify items as substances or mixtures.
		LP4—Identify characteristics as properties or nonproperties.
		LP5—Design an investigation to determine whether two items are the same substance. Make a prediction, identify variables, control variables, and communicate scientific procedures.
		LP6—Conduct a scientific investigation to gather data about properties of substances, such as color, hardness, density, melting point, and solubility.
		LP7—Analyze and interpret data about properties of substances to identify the substance and determine whether two items are the same substance or different substances.
		LP8—Construct a scientific explanation that includes a claim about whether two items are the same substance or different substances, evidence in the form of properties of the substances, and reasoning that different substances have different properties.

When substances interact to form new substances, the elements composing them combine in new ways. In such recombinations, the properties of the new combinations may be very different from those of the old. (AAAS 1990, p. 47)	Substances have distinct properties and are made of one material throughout. A chemical reaction is a process where new substances are made from old substances. One type of chemical reaction is when two substances are mixed together and they interact to form new substance(s). The properties of the new substance(s) are different from the old substance(s). When scientists talk about "old" substances that interact in the chemical reaction, they call them reactants. When scientists talk about new substances that are produced by the chemical reaction, they call them products.	LP9—Define a *chemical reaction* as a process in which two or more substances interact to form new substances with different properties from the old substances.
		LP10—Identify processes as chemical reactions, phase changes, or mixing.
		LP11—Design an investigation to determine whether a process is a chemical reaction. Make a prediction, identify variables, control variables, and communicate scientific procedures.
		LP12—Conduct a scientific investigation to gather data about properties of substances before and after a process (chemical reaction, phase change, mixing).
		LP13—Construct a scientific explanation that includes a claim about whether a process is a chemical reaction, evidence in the form of properties of the substances and/or signs of a reaction, and reasoning that a chemical reaction is a process in which substances interact to form new substances so that there are different substances with different properties before, as compared to after, the reaction.

*National Research Council (NRC). 1996. National science education standards. *Washington, DC: National Academy Press.*

*American Association for the Advancement of Science (AAAS). 1990. Science for all Americans. *New York: Oxford University Press.*

formances that represent the understanding and inquiry learning behind those ideas. Our learning performances delineate multiple ways for students to demonstrate their understanding of the science content in a standard, such as thoroughly defining scientific concepts, designing and conducting investigations, analyzing and interpreting data, and constructing scientific explanations (see Table 1). The learning performances were carefully selected, planned, and organized to provide sequential and coherent experiences with the ambitious content learning goals.

The third step involved developing assessment tools that aligned with our learning performances. To accomplish this, we began by developing a set of base rubrics that corresponded to the different cognitive processes articulated in our learning performances (e.g., define, identify, design, analyze and interpret, and explain). A base rubric describes the different components of a particular way of knowing and the levels of those components. It is a template from which highly specified rubrics can be developed for assessing students on learning performances. For instance, a base rubric for scientific explanation identifies the key components of a scientific explanation (i.e., claim, evidence, and reasoning) and the general performance levels for each component (see Table 2). Because the base rubric for explanation is generic in design and not tied to content, it can be tailored for use to evaluate students' explanations across units, whether students are writing explanations in chemistry, biology, or physics. In this way, the base rubrics encourage greater alignment across units by emphasizing scientific practices that are consistently evaluated with the same criteria. Although we focus exclusively in this chapter on our chemistry unit, "New Stuff From Old Stuff," we are developing a comprehensive middle school curriculum that spans grades 6–8. Our base rubrics will serve to encourage alignment across all of our units.

Our final task in step three was to use our base rubrics to develop specific rubrics for assessing students on each learning performance for our chemistry unit. A specific rubric has the same components and levels as a base rubric but is tailored to a given learning performance. Because specific rubrics directly align with learning performances, they are content specific and can only be used for a certain science topic. Table 3 presents a specific rubric for assessing students' explanations for learning performance 13 of "New Stuff From Old Stuff" (see LP13 in Table 1).

In our fourth step, we used the specific rubrics as a tool to structure assessment and learning tasks for "New Stuff From Old Stuff." In this way, students were evaluated consistently across the unit, using the criteria of the specific rubrics and thereby facilitating alignment between important science content and inquiry standards. The tasks we designed provided multiple ways for students to engage in the science content and demonstrate their understanding.

As a way of contextualizing the content in real-world student experiences, we organized the learning tasks into a series of lessons linked together by a driving question. This was both the fifth and sixth steps in our design process. A driving question uses everyday language to connect with students' authentic interests and curiosities about the world (Krajcik, Czerniak, and Berger 2003). It is carefully crafted as the central organizing feature that drives students' investigations during an inquiry unit. The driving question of our chemistry unit—"How can I make new stuff from old stuff?"—addresses how new substances can be made from old substances. Specifically, students investigate how soap can be made from lard and sodium hydroxide. During the unit, students complete a number of investigations, each time cycling back to the driving question. The investigations allow them to experience scientific phenomena and processes by describing observations, designing and conducting experiments, gathering and analyzing data, and explaining scientific ideas. Each cycle helps students delve deeper into the science content to initially understand substances, then properties, and finally substances interacting to form new substances (i.e., chemical reactions).

EXPLANATION AS A WAY OF KNOWING

Explanation is both a process in scientific inquiry and an important scientific practice (NRC 1996). A significant body of research treats explanation as a process of

TABLE 2.

Base Explanation Rubric

Component	Level		
	1	2	3
Claim An assertion or conclusion that answers the original question.	Does not make a claim, or makes an inaccurate claim.	Makes an accurate but incomplete claim.	Makes an accurate and complete claim.
Evidence Scientific data that supports the claim. The data need to be appropriate and sufficient to support the claim.	Does not provide evidence, or only provides inappropriate evidence (evidence that does not support claim).	Provides appropriate, but insufficient evidence to support claim. May include some inappropriate evidence.	Provides appropriate and sufficient evidence to support claim.
Reasoning A justification that links the claim and evidence and shows why the data count as evidence to support the claim by using the appropriate and sufficient scientific principles.	Does not provide reasoning, or only provides reasoning that does not link evidence to claim.	Provides reasoning that links the claim and evidence. Repeats the evidence and/or includes some scientific principles, but not sufficient.	Provides reasoning that links evidence to claim. Includes appropriate and sufficient scientific principles.

coordinating evidence and theory. The use of explanation can provide students with opportunities to develop competency in this scientific practice (Driver, Newton, and Osborne 2000; Kuhn 1993; Sandoval 2003). Science education researchers have examined students' explanations for the insights they provide into students' understanding of concepts, such as gears (Metz 1991), natural selection (Sandoval 2003), and light (Bell and Linn 2000). These researchers share the view that explanation is more than a simple index of content knowledge; explanation is a key learning goal for science instruction.

The base rubric for explanation entails three components: a claim about a problem, evidence for the claim, and reasoning that links the evidence to the claim (see Table 2). The rubric defines a range of levels for completing each component of an explanation task and can be applied across science content. We applied the base explanation rubric to learning performance 13 in Table 1. Table 3 presents the specific rubric for assessing students' explanations relevant to this learning performance. The three components of the explanation rubric allowed us to analyze separately a student's claim, evidence for the claim, and reasoning linking evidence and claim.

USING THE EXPLANATION RUBRIC IN THE CLASSROOM

One of our goals was to create usable assessment tools for classrooms. Teachers can customize the base rubric for explanation to create a specific explanation rubric for any grade level or science content area. Teachers can chart the progress of their students' explanations by using the same rubric for assessment tasks throughout a unit or entire curriculum. They can adapt the base rubrics for the learning goals and student ability levels in their classrooms. In order to customize the rubrics, a teacher would need to go through a process of learning-goals-driven design similar to the one described previously in this chapter. The teacher needs to determine the learning goals of the unit, then align the learning goals and the base rubrics to create specific rubrics.

For example, "New Stuff From Old Stuff" requires students to explain whether or not a chemical reaction occurred after different substances were mixed together (learning performance 13 in Table 1). The specific explanation rubric in Table 3 breaks down the explanation into the three components and the levels of response for each component. First students make a *claim*: an assertion about what they think is happening. After observing a chemical reaction, a student might claim, "A new substance was formed after mixing substances together." This claim is accurate and would be rated a level 3 response (see Table 3). Next, students provide *evidence*: scientific data to support their claim. For example, a student could write, "This new substance has a different density, color, and melting point than the substances I started with." Because this student provides three appropriate pieces of evidence, the response would be rated a level 3. Note that level 2 for the evidence component is divided into level 2a and 2b, allowing the teacher or researcher to differentiate between students' responses that provide only one or two pieces of evidence. Splitting level 2 into 2a and 2b illustrates further how a specific explanation rubric may be adapted for a particular learning performance. Finally, students provide their *reasoning*: a justification for why the evidence supports their claim using the appropriate and sufficient scientific principles. For example, a student could write, "Because there are new properties (density, color, and melting point), I know there is a new substance. Different substances have different properties." This reasoning would be rated a level 3. Separating a student's explanation into these three components and considering the level of response for each component can provide greater insight into the evaluation of student understanding.

To illustrate how the specific explanation rubric can be used, we describe in the next section how we used the rubric to assess student learning during the "New Stuff From Old Stuff" unit in an urban science classroom.

TABLE 3.

Specific Explanation Rubric for Learning Performance 13 (LP13)*

Component	Level 1	Level 2		Level 3
Claim An assertion or conclusion that answers the original question.	No claim, or an inaccurate claim. Sample Response: *"Something happened after mixing."*	[Does not apply.]		An accurate and complete claim. Sample Response: *"A new substance was formed after mixing substances together."*

Evidence Scientific data that support the claim. The data needs to be appropriate and sufficient to support the claim.	Does not provide evidence, or provides evidence that does not support the claim.
	
	Sample Response: *"When the substances were mixed together, I saw that they changed."*
		(a) Provides one piece of appropriate evidence. May include some evidence that does not support the claim.
		
		Sample Response: *"When the substances were mixed together, I noticed that it made more stuff and changed color."*
			(b) Provides two pieces of appropriate evidence. May include some evidence that does not support the claim.
			
			Sample Response: *"When the substances were mixed together, I saw that a gas formed and that it changed color."*
				Provides three of the following pieces of appropriate evidence, if applicable:
				  changes color
				  changes hardness
				  changes odor
				  changes melting point
				  changes solubility properties
				  different density
				  different pH
				  produces a gas ("bubbles"; "fizzes")
				  produces a solid or precipitate
				  produces heat
				
				Sample Response: *"When the substances were mixed together, I noticed that the mixture bubbled, changed color, and became hot."*

Reasoning A justification that links the claim and evidence and shows why the data count as evidence to support the claim by using the appropriate and sufficient scientific principles.	Does not provide reasoning, or provides reasoning that does not link evidence to the claim. Sample Response: *"Because my evidence shows that a new substance was formed."*	Reasoning includes: Evidence shows that there are new properties after mixing. OR Evidence shows that a new substance was formed after mixing. Sample Response: *"A new substance was formed because my evidence shows that the substances before mixing are different from the substances after mixing."*		Reasoning includes: Evidence shows that there are new properties after mixing. These new properties show that a new substance was formed. Sample Response: *"A new substance was formed because my evidence shows that the properties after mixing the substances were different from before mixing them."*

*LP13: "Construct a scientific explanation that includes a claim about whether a process is a chemical reaction, evidence in the form of properties of the substances and/or signs of a reaction, and reasoning that a chemical reaction is a process in which substances interact to form new substances so that there are different substances with different properties before compared to after the reaction."

RESEARCH QUESTION AND METHODS

We field-tested "New Stuff From Old Stuff" during the 2001–2002 school year to evaluate usability for teaching and effectiveness for student learning. We were particularly interested in investigating the research question: In what ways do our base and specific rubrics, aligned with curriculum and national standards, enable assessment of student learning throughout instruction? The setting for our research was a public middle school located in a large urban school district in the Midwest. We worked with an experienced science teacher, Katheryne Frank, and her seventh-grade science class.

MS. FRANK'S CLASSROOM

In spring 2002, Katheryne Frank implemented "New Stuff From Old Stuff" with her seventh-grade science class. Ms. Frank was an experienced science teacher with familiarity in using inquiry-based instructional materials. Prior to teaching our unit, her curriculum had focused on scientific process skills, the Earth's atmosphere, and the composition of matter, including the states of matter and the particulate nature of matter. Thirty-two students were enrolled in her class, although attendance ranged from 19 to 27 students on any one day. The school, typical compared with other middle schools in the district, was made up of approximately 470 students, primarily African Americans from lower to lower-middle income families. Nearly all students qualified for free or reduced-price lunch.

DATA COLLECTION

"New Stuff From Old Stuff" focused on increasing student understanding of two central ideas in chemistry—substances and properties, and the nature of chemical reactions. Students were given written pre- and post-content assessments (pretest-posttest design). The pre- and post-assessments were identical and consisted of a total of 20 multiple-choice and 4 open-ended items. Open-ended items were assessed using specific rubrics. In addition to pretest-posttest measures, students' written assignments during the unit were collected as artifacts and assessed using specific rubrics. Artifacts included work from students' science folders, such as written explanations based on their observations of scientific phenomena and their claims regarding experiments they had conducted. Observers were present in the classroom daily during the unit, taking field notes and videotaping class sessions.

DATA ANALYSIS

We developed specific rubrics for assessing students' written explanations on pre- and posttest open-ended items and student artifacts from the unit. A specific rubric defined a range of levels from 1 (low) to 3 (high) for completing each component of an explanation. Three raters independently assigned ratings on pretest and posttest open-ended items for all students. The raters agreed on 91% of level ratings; disagreements were resolved through discussion. Raters assigned level ratings through discussion and then agreement by consensus on the written explanations collected as artifacts during the unit.

FINDINGS

As mentioned earlier, one of our goals was to create usable assessment tools for teachers and researchers. During the enactment of "New Stuff From Old Stuff" in Ms. Frank's classroom, we put our specific explanation rubric to use for the purpose of looking at how learning progressed over the course of the unit. In order to provide a picture of how students' explanations changed, we focus in this section on three students: Elena, Asha, and Bethany (names of students are pseudonyms). The three students reflected different levels of ability in writing explanations at the beginning of the unit and each showed unique growth. On occasion, we also refer to examples from other students. To illustrate how the explanation rubric was used for ongoing assessment, we first examine students' written explanations from the pretest and beginning of the unit. Then we describe how Ms. Frank worked with her students to help them improve the quality of their explanations. Finally, we provide a description of students' explanations at the end of the unit and consider their progress from beginning to end.

STUDENT EXPLANATIONS ON THE PRETEST AND BEGINNING OF THE UNIT

Using a rubric early in a unit can help teachers and researchers plan instruction to meet the particular needs of students. To illustrate this possibility, we applied the specific explanation rubric in Table 3 to students' explanations of a chemical reaction prior to starting "New Stuff From Old Stuff." The students completed a pretest that included open-ended problems requiring explanation. The following problem dealt with a chemical reaction:

1. *You have a clear liquid, a white powder, and a red powder.*
2. *Design an experiment to find out which two substances when mixed together will produce a chemical reaction.*
3. *Describe three pieces of evidence you would look for to determine if a chemical reaction occurred.*
4. *Why does this evidence support that a chemical reaction occurred?*

We focused on students' responses to the latter two steps of the problem, because they correspond to the evidence and reasoning components of our explanation rubric. The claim was given: A chemical reaction occurred. The second step required students to provide evidence for the claim and the third step required students to provide reasoning linking that evidence to the claim. The evidence and reasoning of our three target students on the pretest, and for the same task on the posttest, is found in Table 4.

TABLE 4.
Pretest and Posttest Written Responses of Elena, Asha, and Bethany*

Open-Ended Item: Design an experiment to find out which two substances when mixed together will produce a chemical reaction. Describe three pieces of evidence you would look for to determine if a chemical reaction occurred. Why does this evidence support that a chemical reaction occurred?

Component	Test	Elena's Responses	Asha's Responses	Bethany's Responses
Evidence	Pretest	No response. (Level 1)	"The substances will have reacted, you will see a change, and you will just compare." (Level 1)	"A color change bubbling." (Level 2b)
	Posttest	"Change in color, or bubbles, powder dissolving." (Level 2b)	"The color, the hardness, and the density is evidence I would look for. Because you could look for the properties and see if they changed." (Level 3)	"bubbleing," "color change," "Density" (Level 2b)
Reasoning	Pretest	No response. (Level 1)	"This evidence supports that a chemical reaction occurred because you can follow the evidence and determine that it change." (Level 1)	"It will because it change of formed." (Level 1)
	Posttest	"Because it is A it is A chemical change because new things happen." (Level 1)	"Because you could look for the properties and see if they changed." (Level 2)	"because when you mix two things they Just don't bubble it have to be something th-[illegible]." (Level 1)

*Students' original spelling, grammar, and punctuation are intact in all examples.

Asha's explanation was representative of her classmates' on the pretest. Her response lacked enough detail to count as evidence according to the specific explanation rubric. She wrote in general terms ("you will see a change") rather than citing particular pieces of evidence of a chemical reaction, such as a change in color or the production of a gas. Asha's evidence received a level 1 rating (ratings of levels in this section refer to those in Table 3). Although Asha referred to both her evidence and the claim when asked to provide reasoning, she did not provide a justification articulating *why* her evidence supported the claim, nor did she include appropriate scientific principles. Her attempt at reasoning was rated a level 1.

Bethany provided a somewhat different explanation than did Asha by including two pieces of plausible evidence of a chemical reaction (her evidence was rated as level 2b). Bethany was one of only a few students who provided appropriate evidence on the pretest. On the other hand, Bethany, like Asha, was unable to articulate a reason why her evidence supported the claim. It was not clear what she thought her evidence accomplished. In fact, reasoning was an impediment for all students on the pretest. Other examples of students' reasoning for this question included: "Cause that is what happen when a chemical reaction occurs." "It would no longer be a liquid." and "By just reacting the substances they had being mixed together." Each of these attempts was rated level 1.

Elena did not answer the open-ended problem about a chemical reaction on her pretest, although she completed other questions. To characterize her explanations at the beginning of the unit, we examined her first opportunity to provide an explanation within the unit. In the first lesson of "New Stuff From Old Stuff," students recorded descriptions of two "unknown" materials. The students wrote a claim stating whether the two unknowns were the same or different and provided evidence for their claim. The explanation task corresponded to the claim and evidence components of the base explanation rubric. Elena claimed that the two unknown materials "are the same," a perfectly reasonable claim, given that the materials were unknown. Yet providing appropriate evidence for a claim was a challenge; her evidence was that they "both leave grease spots," even though she recorded that only a single unknown was "greasy" in her descriptions. To support her claim that they "are the same," Elena needed to select appropriate evidence such as two similar characteristics of the unknown substances. She did not yet understand what counted as appropriate evidence.

In general, Ms. Frank's students made claims in their explanations at the beginning of the unit. The explanation rubric revealed that most students had difficulty providing evidence to support their claims, due to either inappropriateness, as exemplified by Elena, or a lack of detail, as exemplified by Asha. Most students' evidence was rated as level 1, with Bethany the rare exception. The rubric, when applied early during the unit, also showed that Ms. Frank's students had difficulty providing reasoning linking their evidence and claims. Their early attempts at reasoning were poorly articulated and scored as level 1.

ADDRESSING STUDENT DIFFICULTIES THROUGH CLASSROOM DISCUSSIONS ABOUT EVIDENCE

On the fourth day of the chemistry unit, Ms. Frank used the term *evidence* for her bell work—what she called the "science thought" of the day. Ms. Frank selected and emphasized this term from the unit materials because she was concerned that her students would not understand how to use evidence in their explanations. The students created a diagram for the word *evidence* with four components: define what is meant by evidence, describe what evidence is used for, give examples of evidence, and provide "nonexamples" of evidence (Ms. Frank's term for a counter example). Ms. Frank led a discussion about their diagrams of *evidence*.

Ms. Frank:	*Evidence. What does it mean?*
Student:	*It shows what is true and what is not.*
Ms. Frank:	*How have we used evidence this week?*
Student:	*To observe things.*
.....	
Ms. Frank:	*You use evidence, OK. Not just in a courtroom, you use evidence when you want to say something.... "How do you know" is your evidence.... What is a nonexample?*
Student:	*An opinion.*
Ms. Frank:	*That is a pretty good nonexample. Some people might think that just because they believe something that that's evidence. But the evidence is why you believe something. You've got to have reasons for believing something. That's your evidence.*

On the next day, Ms. Frank approached one of the researchers before class. Ms. Frank explained that, while reviewing students' work from previous lessons, she noticed that her students were having difficulty describing observations with appropriate evidence. During a lesson in which students were given "unknown" materials and asked to carefully observe and write descriptions with appropriate evidence, they often overlooked the physical evidence in front of them. Ms. Frank commented that students were naming the materials (e.g. "looks like cotton candy") and giving their opinions (e.g. "it could be used for cooking chicken") instead of recording observable properties such as color and hardness.

To address the students' confusion, Ms. Frank gave her students the vocabulary word *observe* for their science thought that day. After she led a discussion of the students' diagrams for their science thought, she addressed a number of the students' difficulties. She told her students that she had read their descriptions of the unknown materials and that some of what they wrote were "not really observations."

| Ms. Frank: | *Some of you wrote, "It's soap." Now if you wrote, "It smells like soap," that was OK because that's an observation based on prior knowledge. But if you told me, "It was soap" that is not an observation. That is an opinion because I haven't told you what it was. You just looked at it.* |

By discussing *evidence* and *observe*, Ms. Frank helped her students understand what they needed to include in explanations. One technique Ms. Frank used was to remind students to "CQAA"—Combine Question And Answer. This technique encouraged students to think about what the question was asking and to include the question in their response. CQAA correlated with the concept of *claim* in the explanation rubric. Ms. Frank's use of CQAA is one example of how a teacher might customize the explanation rubric to match with one's own instructional strategies and the unique needs of students.

Because this technique was a part of the classroom culture prior to the start of the unit, it may be one reason why students consistently wrote appropriate claims even on the pretest. Because the use of evidence and reasoning in the students' explanations was new, it was not surprising that students had difficulty with the explanation components at the beginning of the unit. As students became accustomed to being asked for evidence, and after Ms. Frank's discussion of evidence, we saw improvement in the evidence component of the students' explanations.

STUDENTS' EXPLANATIONS AT THE END OF THE INSTRUCTIONAL UNIT

A rubric can provide detail about the progress of individual students. Comparing student work at different times (beginning, middle, or end of a unit) can also help identify trends in student thinking and uncover common student difficulties. We applied the specific explanation rubric to student explanations from the last lesson and the posttest.

The culminating lesson

In the final lesson, students conducted experiments in which they made soap from lard and sodium hydroxide. Afterwards, students wrote claims about whether they thought a new substance was formed as a result of their experiments, provided evidence to support their claims, and articulated their reasoning (Table 5). Asha, Bethany, and Elena made correct claims that a new substance was formed; their responses were scored a level 3. Additionally, all were able to provide some appropriate evidence. Their explanations varied most on the reasoning component.

Asha's evidence included two properties that changed (color and odor). While she also stated "hardness," there was no mention of change (not counted as appropriate evidence). For this reason, the evidence component rated a level 2b. Her reasoning statement received a level 2 because she mentioned that her evidence supported a change in substances, but she did not establish that this was due to a change in properties.

Bethany stated, "The Color went form off white to milkey white. The Hardness went form soft squishy to seim [semi] hard." Her statement summarized two changes in properties and was rated level 2b. Bethany's reasoning was not characteristic of her classmates. She was the only student to provide a justification that linked her evidence (changes in properties) to her claim (make a different substance).

Elena's response included three appropriate properties (hardness, solubility, and color) so she received a level 3 on the evidence part of the rubric. She also included "it is a new substance" as a piece of evidence, perhaps simply a restatement of her claim, or suggesting that she was unclear about what counted as evidence. Her reasoning statement received a level 1 because she did not include a justification linking the properties of the substances to her claim that a new substance was formed.

TABLE 5.

Written Responses of Elena, Asha, and Bethany in the Culminating Lesson of "New Stuff From Old Stuff"*

Q: Do you think a new substance was formed after mixing the fat, the rubbing alcohol, and sodium hydroxide? Provide 3 pieces of evidence to support your answer. Explain why the evidence supports your answer.

Component	Elena	Asha	Bethany
Claim	"it is a new substance… because I don't see any salt, or alcohol, or sodium hydroxide." (Level 3)	"a new substance was formed after mixing the fat, the rubbing alcohol and sodium hydroxide a chemical change." (Level 3)	"Yes because when they were not combined they was Just stuff some liquid some not but when they were mixed they made A hard or ['soft' or 'solid'—unclear] type of soap." (Level 3)
Evidence	"1. it is a new substance. 2. it has new properties. 3. <u>fat</u> hardness: soft squishy solubility: water – no oil – yes color: off white <u>soap</u> hardness: semi-hard solubility: water – yes oil – no color: milky white" (Level 3)	"3 pieces of evidence to support your answer is color (The color went from off white to milky white), smell (It went from stinky to a no smell), and the hardness." (Level 2b)	"The sodium hydroxide was [incomplete]… The salt was clear. The rubbing alcohol was clear. After you combine they the became A hard whit piece of soap. The Color went form off white to milkey white. The Hardness went form soft squishy to seim [semi] hard." (Level 2b)
Reasoning	"Because it shows what each substance has in it." (Level 1)	"the evidence support my answer because you could do the experiment and see that the same changes will occur and you will get a new substance from the other substances." (Level 2)	"It support my Answer because you see that After combining everything you get Different Color, Hardness, Density, Solubility, and ph. These Are changes and properties which make a different substance." (Level 3)

*Students' original spelling, grammar, and punctuation are intact in all examples.

Students' explanations on the posttest

The students took the same pre- and posttest (see Table 4). Asha's evidence improved considerably. Where her pretest response did not include enough detail to count as evidence and received a level 1 rating, her posttest response included three pieces of appropriate evidence and was rated a level 3. Her reasoning score also improved. She received a level 2 on the posttest for talking about a change in properties.

Bethany's explanation for the posttest problem differed slightly from her pretest response. The evidence component of her explanation included two pieces of appropriate evidence (bubbling and color change). While she stated "density," there was no mention of change. For this reason, her evidence reflected a level 2b. Bethany's reasoning on the posttest did not improve compared to the pretest. In both cases, her reasoning was rated a level 1.

While Elena did not respond to the open-ended questions on the pretest, her evidence on the posttest was considerably stronger than her evidence from the first lesson of the unit. Elena's posttest included two pieces of appropriate evidence (change in color, bubbles), so her response was classified as Level 2b. She also mentioned "dissolving" (which was not evidence for a chemical reaction according to the specific rubric). By using the rubric to analyze student explanations, we were alerted to a number of students who shared Elena's misconception that dissolving or mixing is a chemical reaction. Elena's reasoning received a level 1 because she did not provide a justification for why her evidence supported her claim.

CHANGES IN STUDENTS' EXPLANATIONS

By the end of the unit, the three target students improved their abilities to construct explanations. Elena improved in selecting evidence to support her claims, but her reasoning statements remained low in quality on both the culminating lesson and posttest. The development of Elena's explanations was characteristic of the majority of the students in the class. Asha improved the quality of the evidence for her claims by providing more detail. She also showed some improvement in reasoning by providing scientific principles to link her claim and evidence. Bethany was already adept at noting evidence at the beginning of the unit and was the only student to give a high-level reasoning statement on the culminating lesson. However, her reasoning statement on the posttest did not reflect this improvement.

CONCLUSIONS

We expected students to articulate their reasoning in their explanations by the end of the unit. At first, we were surprised that they did not do this; however, when we reviewed our unit materials, we realized that the materials did not communicate to the students or teacher what counted as reasoning or explicitly state that students should include reasoning. It was not surprising that students' responses scored low for reasoning.

While the first six steps of our learning-goals-driven design process resulted in greater alignment between the curriculum and the assessments, the seventh step in our design process—the pilot testing of the materials in the real-world environment of a science classroom—informed us of the gap between our expectations and the actual unit. Similarly, the materials did not communicate to the teacher or students what should count as evidence. Ms. Frank specifically discussed with her students what was meant by scientific evidence and what counted as scientific evidence. When we used the specific explanation rubric to evaluate students' responses from the last lesson and posttest, we found that students scored high in their use of evidence. Using this and other data from our field test, we revised both the assessments and the activities in another cycle of learning-goals-driven design (McNeill et al. 2003).

Assessments that tackle multiple ways of knowing with many components yield valuable information about student learning. A wide range of knowledge and understanding can arise from instruction, but student competencies and difficulties may not be revealed by assessments that treat knowledge as unitary. By parsing explanation as a way of knowing into three components, we found that students became skilled at two of the components (making claims and providing evidence) but remained challenged by the third component (reasoning). Assessments that incorporate a range of knowledge are likely to be more informative for teachers by further clarifying what it means for students to "know" science.

Our work suggests that this kind of assessment could be useful for teachers and researchers to follow the evolution of student learning across grade levels or science content areas. The base rubric and its derivative specific rubrics were used to compare students' explanations at the beginning and end of the unit. The same base rubric could be applied to students' explanations in a subsequent science unit or even a different grade level.

A strength of base rubrics, aligned with curriculum and important learning standards, is that a teacher or researcher can customize them for classroom use and obtain evidence of student learning immediately. Measures of student learning that are more distant from the standards and from the daily learning experiences of students cannot provide such data. Standardized tests have become the gold standard and have become a high priority for political and policy decisions, but they are not as immediately helpful for judging students' understanding. Indeed, results of large-scale standardized exams are rarely reported in a timely enough manner for practical and informative classroom use. In contrast, base rubrics and other forms of usable assessments enable teachers to take an active role in the analysis of student work. Such assessments can help reveal instructional gaps and suggest courses of action for teachers, allowing teachers to shift instruction (based on assessment), and support learning as students progress through a unit of study. This, we believe, holds tremendous promise for narrowing the divide between assessment and instruction, and ensuring that students meet the educational standards called for in science education reform.

ACKNOWLEDGMENTS

We gratefully acknowledge the collaboration of Katheryne Frank and her students. The research reported here was supported in part by the National Science Foundation (REC 0101780 and 0830310A605). All opinions expressed in this work are the authors' and do not necessarily represent either the funding agencies or the authors' institutions.

LINKS TO THE NATIONAL SCIENCE EDUCATION STANDARDS

In this chapter, we reported on how the learning-goals-driven design process used the National Science Education Standards (NSES) as the starting point for the alignment of assessment, curriculum materials, and instruction for middle school inquiry science. The assessment standards of the NSES emphasize changes in assessment practice (NRC 1996, p. 100), which was consistent with the goals of our work. Science as Inquiry Content Standard A 5–8 and Physical Science Content Standard B 5–8 provided a focus for our alignment process.

REFERENCES

American Association for the Advancement of Science (AAAS). 1990. *Science for all Americans: A Project 2061 report on literacy goals in science, mathematics, and technology.* New York: Oxford University Press.

American Association for the Advancement of Science (AAAS). 1993. *Benchmarks for science literacy.* New York: Oxford University Press.

Bell, P., and M. Linn. 2000. Scientific arguments as learning artifacts: Designing for learning from the web with KIE. *International Journal of Science Education* 22 (8): 797–817.

Bransford, J. D., A. L. Brown, and R. R. Cocking, eds. 2000. *How people learn: Brain, mind, experience, and school.* Washington, DC: National Academy Press.

Driver, R., P. Newton, and J. Osborne. 2000. Establishing the norms of scientific argumentation in classrooms. *Science Education* 84 (3): 287–312.

Krajcik, J., C. M. Czernizk, and C. F. Berger. 2002. *Teaching science in elementary and middle school classrooms: A project-based approach.* 2nd ed. New York: McGraw Hill.

Kuhn, D. 1993. Science as argument: Implications for teaching and learning scientific thinking. *Science Education* 77 (3): 319–337.

McNeill, K. L., D. J. Lizotte, C. J. Harris, L. A. Scott, J. Krajcik, and R. W. Marx. 2003. Using backward design to create standards-based middle-school inquiry-oriented chemistry curriculum and assessment materials. Paper presented at the Annual Meeting of the National Association for Research in Science Teaching, Philadelphia, PA (March).

Metz, K. 1991. Development of explanation: Incremental and fundamental change in children's physics knowledge. *Journal of Research in Science Teaching* 28 (9): 785–797.

National Research Council (NRC). 1996. *National science education standards.* Washington, DC: National Academy Press.

Pellegrino, J. W., N. Chudowsky, and R. Glaser, eds. 2001. *Knowing what students know: The science and design of educational assessment.* Washington, DC: National Academy Press.

Perkins, D. 1998. What is understanding? In *Teaching for understanding: Linking research with practice,* ed. M. S. Wiske. San Francisco, CA: Jossey-Bass.

Reiser, B. J., J. Krajcik, E. Moje, and R. Marx. 2003. Design strategies for developing science instructional materials. Paper presented at the Annual Meeting of the National Association for Research in Science Teaching, Philadelphia, PA (March).

Ruiz-Primo, M. A., R. J. Shavelson, L. Hamilton, and S. Klein. 2002. On the evaluation of systemic science education reform: Searching for instructional sensitivity. *Journal of Research in Science Teaching* 39 (5): 369–393.

Sandoval, W. A. 2003. Conceptual and epistemic aspects of students' scientific explanations. *Journal of the Learning Sciences* 12 (1): 5–51.

Shepard, L. A. 2000. The role of assessment in a learning culture. *Educational Researcher* 29 (7): 4–14.

Simon, H. A. 1996. *The sciences of the artificial.* 3rd ed. Cambridge, MA: MIT Press.

Wiggins, G., and J. McTighe. 1998. *Understanding by design.* Alexandria, VA: Association for Supervision and Curriculum Development.

RESOURCES

Arter, J., and J. McTighe. 2001. *Scoring rubrics in the classroom: Using performance criteria for assessing and improving student performance.* Thousands Oaks, CA: Corwin.

A practical guide to effective assessment for improved student learning.

Atkin, M. J., and J. E. Coffey, eds. 2003. *Everyday assessment in the science classroom.* Arlington, VA: NSTA Press.

A collection of essays on supporting learning through the use of assessments that are an integral part of daily classroom activities.

Bransford, J. D., A. L. Brown, and R. R. Cocking, eds. 2000. *How people learn: Brain, mind, experience, and school.* Washington, DC: National Academy Press.

This book examines recent research findings about thinking and learning and considers implications for educational practice.

Brown, J. H., and R. J. Shavelson. 1996. Assessing hands-on science: A teacher's guide to performance assessment. Thousands Oaks, CA: Corwin.

This assessment handbook for science teachers who use hands-on science curricula describes a variety of science performance assessments.

Enger, S. K., and R. E. Yager. 2001. *Assessing student understanding in science: A standards-based handbook.* Thousands Oaks, CA: Corwin.

A comprehensive and practical handbook for assessing science learning linked to the National Science Education Standards.

Investigating and Questioning Our World Through Science and Technology (IQWST) Project (website: *www. hi-ce.org/iqwst*).

IQWST is a collaborative effort between curriculum researchers at the University of Michigan and Northwestern University to develop middle school curriculum and assessment materials that help students to learn important science content that is based on national standards.

Krajcik, J., C. M. Czerniak, and C. F. Berger. 2002. *Teaching science in elementary and middle school classrooms: A project-based approach.* 2nd ed. New York: McGraw Hill.

A thorough introduction to the theory and practice of project-based science, this book is keyed to the National Science Education Standards and includes a chapter on assessing in project-based science classrooms.

Minstrell, J., and E. H. van Zee, eds. 2000. *Inquiring into inquiry learning and teaching in science.* Washington, DC: American Association for the Advancement of Science.

This book explores what is meant by inquiry learning and teaching, provides examples of inquiry practice in science classrooms, and addresses issues that arise in inquiry instruction.

Pellegrino, J. W., N. Chudowsky, and R Glaser, eds. 2001. *Knowing what students know: The science and design of educational assessment.* Washington, DC: National Academy Press.

This book provides an in-depth look at how new knowledge about student learning can inform the development of new kinds of assessments. It presents a research-based approach to assessment of student learning, suggests principles for designing new kinds of assessments, and considers implications for education policy, practice, and research.

Stiggins, R. J. 2001. *Student-involved classroom assessment.* 3rd ed. Columbus, OH: Merrill Prentice Hall.
 This comprehensive book for teachers illustrates how to create classroom assessments for documenting
 and improving student learning.

Wiggins, G., and J. McTighe, 1998. *Understanding by design.* Alexandria, VA: Association for Supervision and
Curriculum Development.
 Intended for teachers, curriculum developers, and administrators, this book offers a practical framework
 for designing curriculum, assessment, and instruction.

AUTHOR AFFILIATIONS

Christopher J. Harris is a new faculty member in teaching and teacher education at the University of Arizona. He has a broad background in science education at the elementary and middle school levels. His research interests include teacher education and student learning in science, as well as the design of science curricula and assessments.

Katherine L. McNeill, a former middle school science teacher, is a doctoral student in science education and is pursuing a master's degree in ecology and evolutionary biology at the University of Michigan. Her research interests include student learning and curriculum design specifically for scientific inquiry skills such as constructing explanations.

David J. Lizotte is a doctoral student in psychology at the University of Michigan. Among his research interests is the cognition underlying students' scientific reasoning.

Ronald W. Marx is dean of education and a professor of educational psychology at the University of Arizona. His research and development activities focus on how to help teachers engage their students in thoughtful learning of science.

Joseph Krajcik, a professor of science education at the University of Michigan, focuses his research on designing science classrooms so that learners engage in finding solutions to meaningful, real-world questions through collaboration and the use of learning technologies.

The authors have a long-standing collaboration through the University of Michigan's Center for Highly Interactive Computing in Education.

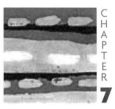

using rubrics to foster meaningful learning

Marcelle A. Siegel, Paul Hynds, Marc Siciliano, and Barbara Nagle

INTRODUCTION

When the school day ended and the last student complaining about a grade finally left the classroom, the middle school science teacher slumped into her chair. How many—if any—students, she wondered, had learned something significant during the unit on household materials? She had posed some short essay questions that asked them to analyze the chemical information on several cleansers and to discuss potential toxic hazards. Students compared the benefits and risks of the use of the materials. When reading the essays, she saw that the students recalled the labels and some of the hazards of certain household chemicals, but could not weigh the advantages and disadvantages of using them. Their responses revealed a lack of critical analysis. She could still hear the student's voice: "What's wrong with my answer? I told you all about the ingredients in bleach and other cleaners and that some of them are dangerous. I should at least get a B!"

Unfortunately, similar scenes are played out in many classrooms. The teacher's expectations are not met by the student's performance, and yet the teacher is not certain what to do next. Students often assume that if they give any kind of detailed answer it should be sufficient. Teachers sometimes find themselves talking with students more about what grades they are getting than about what the students are learning. What kinds of tools would help teachers clarify expectations and focus attention more on learning and producing quality work than on grades?

BACKGROUND

Since its inception in the mid-1980s, the Science Education for Public Understanding Program (SEPUP) at the Lawrence Hall of Science (University of California-Berkeley) has developed an array of "issue-oriented" instructional and assessment materials that meet the recommendations of major reform efforts and the expectations of schools. Students gather scientific evidence during guided investigations and apply the evidence during discussions, debates, role plays, and other activities related to societal or personal issues (Thier and Nagle 1994).

In order to assess students' higher-level thinking, decision-making, and process skills, a new type of assessment system was necessary. In collaboration with the Berkeley Evaluation and Assessment Research (BEAR) group at the University of California at Berkeley Graduate School of Education, SEPUP developed an embedded, authentic assessment system as part of the SEPUP full-year course, Issues, Evidence and You (Roberts, Wilson, and Draney 1997; Wilson and Sloane 2000). The basic assessment system had three components:

* Five variables (see Figure 1) that defined the key domains in which students were expected to make progress during the year
* Actual assessment tasks
* Rubrics used to evaluate student performance on the tasks

Each variable had an associated scoring guide (also called a *rubric*, the term we will be using throughout this chapter), which provided criteria for different levels of student performances. These criteria set the standards of performance for different levels of responses to a task. These levels, or standards of performance, not only informed students of what they were doing, but pointed the way to improving performance by setting clear expectations.

In this chapter, we discuss the use and adaptation of rubrics in many classrooms nationally as teachers (and students) moved toward meeting their goals. One SEPUP teacher wrote about the power of using rubrics: "Students knew what was expected of them and it showed in some of the answers that I received. They thought more thoroughly about what they were going to answer."

Frequent and systematic use of rubrics in the classroom can fundamentally change the dynamics of the teacher-learner interaction. This change can be seen in greater student motivation, improved learning and metacognitive skills, and greater teacher understanding of students' learning. The use of rubrics can also mirror scientific habits of mind and reinforce the importance of evidence.

The SEPUP assessment system was cited by the National Research Council as an exemplary model of measurement (NRC 2001). It was originally developed for a specific middle school course, Issues, Evidence and You. Results from a series of studies indicated that the assessment system was psychometrically sound, that SEPUP students out-performed control students, and that SEPUP users of the

FIGURE 1.
The SEPUP Assessment System Variables

Scientific Process

Designing and Conducting Investigations (DCI)—Designing a scientific experiment, performing laboratory procedures to collect data, recording and organizing data, and analyzing and interpreting the results of an experiment.

Evidence and Tradeoffs (ET)—Identifying objective scientific evidence as well as evaluating the advantages and disadvantages of different possible solutions to a problem based on the available evidence.

Scientific Concepts

Understanding Concepts (UC)—Understanding scientific concepts (such as properties and interactions of materials, energy, or thresholds) in order to apply the relevant scientific concepts to the solution of problems.

Scientific Skills

Communicating Scientific Information (CM)—Organizing and presenting results of an experiment, or explaining the process of gathering evidence and weighing tradeoffs in selecting an effective solution to a problem that is free of technical errors.

Group Interaction (GI)—Developing skill in collaborating with teammates to complete a task (such as a lab experiment), sharing the work of the activity, and contributing ideas to generate solutions to a given problem.

Source: Science Education for Public Understanding Program (SEPUP), Lawrence Hall of Science (University of California-Berkeley).

assessment system out-performed those who used the course alone (Wilson and Sloane 2000). The SEPUP developers collaborated with field-test teachers to adapt the core components of the assessment system for use in modules and two additional full-year courses—Science and Life Issues (middle school life science) and Science and Sustainability (integrated high school science).

RESEARCH QUESTIONS

The most recent research focused on (1) how teachers used the SEPUP assessment system, (2) how the system was modified to retain its power while making it easier to use, and (3) the additional tools that were incorporated into the system.

The first author of this chapter and SEPUP staff engaged in several ongoing studies to refine the variables and associated rubrics for the Science and Life Issues course, the Science and Sustainability course, and 12 shorter modules. The purpose was to refine the rubrics for different instructional goals, develop additional reliable items, and increase the ease of use for students and teachers. This process included discussion and testing among SEPUP developers, two years of field testing in nearly 50 classrooms, and analysis of data. After two rounds of field-test conferences and collection of written feedback from participants, teacher representatives from the field-test centers joined the SEPUP staff for a summer conference and participated in final revision of the embedded items and the rubrics. Research on the life science course involved (a) piloting new multiple-choice and extended items with 600 students, (b) analyzing data (using Rasch modeling with Conquest software and traditional techniques using SPSS software), and (c) refining items for a national field test in 2002–2003. Based on these studies, SEPUP developed more focused variables, simpler language for the rubrics, approximately 100 additional reliable items, and a feedback form.

DEVELOPING AN EMBEDDED ASSESSMENT SYSTEM

The SEPUP assessment system was based on four principles (Wilson and Sloane 2000). First, it was based on a developmental perspective of student learning, with a focus on student progress over the course of a year. The use of the same five variables and rubrics throughout the year reflected this developmental perspective.

Second, the system matched instructional goals. It was focused on understanding important scientific concepts, the processes of scientific investigation and analysis of information, and evidence-based decision making. To ensure a match between curriculum, instruction, and assessment, the instructional materials and assessment tasks were developed at the same time, and the assessment system was part of all field-testing activities.

Third, the system was designed to generate quality evidence. The assessment tasks, methods of measurement, analysis, and reporting needed to be of high technical quality, which meant maintaining standards of fairness, ensuring that results were compared across time and context, and performing traditional studies of validity, reliability, and equity.

Fourth, the system was built on the principles of teacher management and responsibility. The teacher used the assessment evidence to guide the learning process. SEPUP and BEAR involved teachers in all aspects of the development of the assessment system, including (1) developing the tasks and rubrics, (2) collecting and scoring student work, and (3) interpreting the results.

Alternative assessment created new challenges for teachers, such as finding time to score open-ended responses, translating rubric scores to letter grades, and helping students understand that the new form of assessment was intended to guide learning rather than judge student performance. To help teachers manage the new assessment system, SEPUP staff created tools (see p. 96 for the tools) to help them use the system effectively.

FIVE TYPES OF LEARNING ASSESSED IN SEPUP

The SEPUP assessment system measured five types of content and process learning, called "variables," that were central to the instructional materials. The variables were Designing and Conducting Investigations, Evidence and Tradeoffs, Understanding Concepts, Communicating Scientific Information, and Group Interaction (Figure 1). These five variables represented student learning in terms of the core concepts of SEPUP courses based on decision making about societal issues and clarified the conceptual framework for instruction and assessment.

Two SEPUP rubrics

Each SEPUP variable had an associated rubric that set forth the expected levels of performances for students. For example, the rubrics began with a 1, a minimal response teachers often described as "on your way." A more thoughtful response earned a 2, or "almost there." The goal for students was a 3 for a "complete and correct" response. This included scientific and conceptual understanding, the use of evidence in communicating that understanding, and evaluating alternatives. The criteria for 4, or "going beyond," described answers in which students displayed that they were going further in their thinking, such as connecting the specifics of their responses to other ideas. The five rubrics had these four levels of criteria in common, but each also included specific, unique criteria. For example, the Evidence and Tradeoffs (ET) Rubric (Table 1) measured a core goal of SEPUP courses, the ability to use scientific evidence to analyze the advantages and disadvantages of a real-world decision. At Level 3 in that rubric, a student provided the major reasons for or against a decision and supported each with relevant and accurate scientific evidence. Often, students gathered the evidence in the form of data from a hands-on activity; at other times, they obtained evidence from simulations and readings. One SEPUP field-test teacher wrote that using the ET rubric helped middle school students make decisions: "First of all, the questions [i.e., the assessment questions in the curriculum] ask the students to take a stand, make a decision. Typical middle school youngsters like to ride the fence; noncommittal is safe. So all of a sudden they are expected to come up with at least two options. This threw them quite a bit." Eventually, students were able to clarify their viewpoints based on evidence.

Many teachers also found that secondary students could generate a conclusion, but could not explain their reasoning regarding, and judgment of, multiple sources of evidence. Using the ET rubric helped students improve at this type of sophisticated thinking. One new SEPUP teacher commented, "Students actually realize they must provide evidence for their opinions!"

The Communicating Scientific Information (CM) Rubric (Table 2) measured how well students expressed their arguments and/or ideas. This rubric was designed to measure written (e.g., lab report), oral (e.g., presentation), and visual (e.g., poster) student reports. One element of the variable referred to the organization of the response, and the second element involved technical aspects, such as grammar and eye contact.

TABLE 1.
Evidence and Tradeoffs (ET) Rubric for Use With Middle School Science Students

Score	Using Evidence Response uses objective reason(s) based on relevant evidence to argue for or against a choice.	Using Evidence to Make Tradeoffs Recognizes multiple perspectives of issue and explains each perspective using objective reasons, supported by evidence, in order to make a choice.
4	Accomplishes Level 3 AND goes beyond in some significant way, e.g., questioning or justifying the source, validity, and/or quantity of evidence.	Accomplishes Level 3 AND goes beyond in some significant way, e.g., suggesting additional evidence beyond the activity that would further influence choices in specific ways, OR questioning the source, validity, and/or quantity of evidence and explaining how it influences choice.
3	Provides major objective reasons AND supports each with relevant and accurate scientific evidence.	Uses relevant and accurate scientific evidence to weigh the advantages and disadvantages of multiple options, and makes a choice supported by the evidence.
2	Provides some objective reasons AND some supporting evidence, BUT at least one reason is missing and/or part of the evidence is incomplete.	States at least two options AND provides some objective reasons using some relevant evidence BUT reasons or choices are incomplete and/or part of the evidence is missing; OR provides only one complete and accurate perspective.
1	Provides only subjective reasons (opinions) for choice; uses unsupported statements; OR uses inaccurate or irrelevant evidence from the activity.	States at least one perspective BUT only provides subjective reasons and/or uses inaccurate or irrelevant evidence.
0	Missing, illegible, or offers no reasons AND no evidence to support choice made.	Misses information, illegible, or completely lacks reasons and evidence.
X	Student had no opportunity to respond.	

Source: Science Education for Public Understanding Program (SEPUP), Lawrence Hall of Science (University of California-Berkeley).

TABLE 2.

Communicating Scientific Information (CM) Rubric for Use With Middle School Science Students

	Organization	Technical Aspects
Score	Response logically organizes arguments, evidence and/or ideas related to a problem or issue. Ideas are frequently, but not always, organized in the following way: * Introduction * Explanation of procedures * Presentation of relevant evidence * Consideration/ interpretation of the evidence * Conclusion	Response conveys a concept or idea clearly by using the assigned medium appropriately. Possible forms of communication and areas to examine are * *written (e.g., report)*: sentence structure, grammar, spelling, and neatness * *oral (e.g., presentation)*: enunciation, projection, and eye contact * *visual (e.g., poster)*: balance of light, color, size of lettering, and clarity of image
4	Accomplishes Level 3 AND goes beyond in some significant way.	Accomplishes Level 3 AND enhances communication in some significant way. No technical errors.
3	All parts present and logically organized.	Presents response that is clear and easy to understand, with few minor errors.
2	Shows logical order BUT part is missing.	Provides understandable response BUT clarity is missing in places; technical errors may exist BUT do not prevent audience from understanding the message.
1	Lacks logical order OR is missing multiple parts.	Detracts audience from understanding the message with unclear and technical errors.
0	Missing, illegible, or contains no evidence or ideas related to the task.	Misses evidence, illegible, incoherent, or contains no evidence or ideas related to the task.
X	Student had no opportunity to respond.	

Source: Science Education for Public Understanding Program (SEPUP), Lawrence Hall of Science (University of California-Berkeley).

TEACHER TOOLS IN THE ASSESSMENT SYSTEM

Additional components of the SEPUP assessment system were designed to help teachers use the assessments effectively. Assessment blueprints provided a chronological list of course activities with potential assessment opportunities. Link tests and item banks were sets of additional assessment tasks and questions, some closely tied to the curriculum and some transferring items that were used by the teacher to monitor understanding or conduct a summative assessment. Assessment moderation was a process for teachers to compare ideas for scoring student work, deepen their understanding of students' responses, and come to an agreed-upon standard. The SEPUP developers and BEAR researchers used the moderation process with teachers to select exemplars of student work at each scoring level for a variety of tasks. Exemplars provided teachers with a model of how to score student work and a tool for modeling expectations to students.

FINDINGS

Teachers found that even though the assessment system offered new challenges, it was also rewarding to use (Roberts and Wilson 1998). One SEPUP field-test teacher commented that the assessment system gave her and her students "specifics to look and aim for. We had common language for our discussions. The application of the concepts gave both of us a positive way to judge growth." Another field-test teacher remarked on the usefulness of the exemplars: "I was totally impressed with the quality of writing that I received from students.... Students were empowered by the rubric and the exact criteria for which they would be measured. The Level 3 examples given in the teaching materials also enabled students to have a model for excellence." Another teacher said she spent a lot of time giving students feedback on why they received a score and that it paid off in more focused answers: "[I highlighted] in different colors each of the items I was looking for and scored them. (This was very time consuming and I [used] lots of highlighters.) It was worth it, though! The temptation for [students] to just put down words, write for weeks, was somewhat halted since they knew the criteria. (Quality, not quantity is our slogan—along with evidence, not emotion)."

MODIFICATION AND FIELD TESTING

After the collaborative project to develop the assessment system for the Issues, Evidence and You course, we continued to study and refine the variables and associated rubrics for two more courses—Science and Life Issues and Science and Sustainability. This resulted in more focused variables, with fewer sub-variables and somewhat simpler language in the rubrics. Recently, we refined and field-tested three rubrics (Designing Investigations, Analyzing Data, and Evidence and Tradeoffs) for use with 12 supplementary science modules for grades 6–12. Only one rubric was emphasized in each module.

Our process of adapting the rubrics for the modules included changes in content, clarity, and usability. We first discussed the criteria for each level of the rubric and attempted to achieve clear cut-offs between the levels. We scored additional student work to see if the new rubric was effective, and selected exemplars of student work for each level that illustrated the criteria for that level. Based on this evidence, we revised the rubric and analyzed it in light of additional student work.

During this process, we devised a new tool related to rubrics for teachers called a feedback form (Table 3). Teachers often told us that the rubrics were text-heavy and, at first, overwhelming for students. The feedback form was a concise version of a rubric that only listed the criteria for Levels 3 and 4. It was intended to help the teacher introduce rubrics and to show them how to provide feedback to students. It contained space for the teacher to offer written comments to the student about why his or her answer received a particular score. The feedback form was also designed to help teachers organize their comments to students.

TABLE 3.

Feedback Form for Revised Evidence and Tradeoffs Rubric for Modules

Complete and Correct Response	Yes	Almost	No
You use evidence to support a logical interpretation of the data. You evaluate the source, quality, and/or quantity of evidence.			

Comments:

You accomplish the above and go beyond in some significant way, such as: You present a thorough examination of evidence. You connect your ideas with the science concepts learned. You provide an explanation for why alternative ideas were discarded. You provide suggestions for further relevant investigations. You include a diagram or visual to clarify your ideas. Other:	Yes	Almost	No

Comments:

Source: Science Education for Public Understanding Program (SEPUP), Lawrence Hall of Science (University of California-Berkeley).

Each module and its embedded assessments and associated rubric were field-tested by approximately 25 teachers in 10 centers. Selected teachers from each field-test site joined SEPUP staff during a three-day conference and communicated with project staff regularly during the nine months of testing. At the conclusion of the school year, teachers responded to questions about each activity in the module as well as about the module as a whole. This information was collected in an online database that gave SEPUP staff an efficient way to sort and analyze all of the feedback. Another useful data source was the adaptations to the materials and rubrics that teachers made.

As teachers used rubrics to shift student attention from grades to learning, improvements in performance were observed. For example, at the Young Women's Leadership Charter School in Chicago, students were evaluated on individual outcomes each trimester with task-specific rubrics with three common levels: "Not Yet," "Proficient," or "High Performance." Steve Torres, an eighth-grade physical science teacher, elected not to use rubrics or related scoring rubrics during the first trimester, except for the final assessment. In the second and third trimester, he began using rubrics and consistent assessment language related to the rubrics for all assignments. Torres wrote phrases such as "incomplete" at the top of students' assignments, along with sidebar comments on how to improve work. As one might expect, student outcomes appeared to decrease in the second trimester (possibly because students were held accountable to a higher standard than the first trimester). The number of trimester 1 scores for "Not Yet," "Proficient," or "High Performance" were 206, 449, and 104, respectively. Trimester 2 scores were 292, 366, and 73, respectively. With further use of the criteria and rubrics in the third trimester, the average performance increased dramatically (172, 345, and 240, respectively). While no conclusions can be drawn from a single case during one year, the results suggested what many teachers have observed: The more times that students were exposed to rubrics, the more students performed at the highest levels.

SEPUP middle school students in classrooms using the assessment system achieved statistically and educationally significant gains across the five SEPUP variables, including the ability to use evidence and make tradeoffs (Wilson et al. 1995). Wilson and Sloane (2000) reported that gains for students with teachers who received professional development for fully using and implementing the assessment system during the year were 3.46 times greater than those of the control non-SEPUP group and 2.14 times greater than the SEPUP group that received nonassessment professional development (differences between average gains were significant at the .05 level).

IMPLICATIONS FOR TEACHING: HOW TO USE RUBRICS IN THE CLASSROOM

Based on extensive field-test research and ongoing communication with teachers after they implemented the revised, commercial versions of SEPUP materials, we have learned how teachers used the assessment system, and especially the rubrics, to enhance learning. Following are six recommendations that teachers can apply.

1. Create An Effective Learning Environment

All teachers want their classrooms and schools to provide effective learning environments. Most teachers have in mind a picture of a classroom where the focus of attention is on the exchange of ideas, not on performance rankings (grades). Yet students often do not perceive the classroom in this way. Throughout most students' lives, they have been trained to measure their successes by the grades that they receive. Getting the "right answer" is what counts. How can a teacher create an environment that will help students move beyond the comfort zone of the easy and the familiar into the more challenging, and sometimes threatening, arena of testing their ideas and thinking for themselves?

Because rubrics define the expected levels of student performance, they focus attention on criteria for improvement. For example, in some SEPUP rubrics, the criteria for Level 4 ("going beyond") clarified ways to improve critical thinking, such as questioning the quantity or quality of evidence or making specific suggestions for further investigations. When students internalized this way of analyzing their own work, they began asking, "How can I make this better? What else do I know that is relevant to this problem?" Thus, the process of teaching students to use the rubrics as guides for their work moved the classroom focus to the quality of work—exactly where teachers want to go.

Rubrics foster some of the same habits of mind that educators wish to promote in science students. Accepting a proposition as valid requires the evaluation of information and evidence, rather than individual authority. SEPUP teachers often used the Evidence and Tradeoffs Rubric to help students understand that scientific evidence can provide information for personal and societal decisions and that the interpretation of evidence is an integral part of many human endeavors. Another aspect of the scientific approach fostered by rubrics is that the information available may not yield a "perfect" scientific answer or decision—one must develop an ability to make progress despite uncertainty and to be willing to revise a solution in the light of evidence. In the real world, in science, and in the rubric-based classroom, answers are subject to revision as more evidence is collected. Using the Evidence and Tradeoffs Rubric, students learned about weighing the quality and quantity of evidence on different sides of an issue. Classroom conversations revolved around topics of reliability and reproducibility. Another study also found that SEPUP high school students' attitudes about the relevance of science in their lives were enhanced (Siegel and Ranney 2003).

2. Introduce Rubrics Right Away

All teachers understand the importance of the critical first weeks of a school year. It is during this time that classroom structure is established, goals are set, and expectations declared. In order to implement rubrics successfully, one teacher developed a strategy that focused on building observation skills while introducing a rubric-based assessment system at the same time. Mike Lach, a National Board Certified teacher, designed a unit on observations and interpretations for an introductory unit to physical science courses. One objective of this unit was to help students distinguish between observation and inference. He and Marc Siciliano implemented this unit while teaching ninth grade at Lake View High School in Chicago.

The unit consisted of a series of activities that took students through progressive stages of learning about observations and creating practice rubrics. First, students learned the importance of making detailed observations in science. Students were presented with open-ended stories, ambiguously written, so that multiple interpretations were possible. Students shared their interpretations and saw that confusion existed when stories lacked detail and facts. As the notion of details and facts was expanded, students learned the basic definition of a scientific observation.

Second, the goal was to have students see the value in rubrics by doing the same task twice—once "blind" and the second time with a rubric—and ending with a more coherent outcome. Without a rubric, students were asked to classify a set of observations of an object (a common chalkboard works nicely) to test their understandings of "good" observations. The students initially placed the observations into two groups, one representing "better" observations and the other indicating "poorer" ones. These two groups were then further divided so that there were four groups representing a range from best to worst. Students then named each category and listed the defining characteristics of each. They suggested characteristics such as complete sentences, multiple adjectives, spelling errors, exactness, inclusion of interpretations, and so forth. When the class then scored their observations based on these insufficient categories, they saw that they scored many observations differently. They saw that some observations could be placed in multiple categories and began to recognize the notion of a continuum. These realizations sparked interesting questions about the process and revealed a practical need for rubrics to set criteria for better responses.

As students shared their categories and the rationales for the categories, they saw the limitation of the activity: The task was vague. Thus, the third step was to have students come to consensus by deciding on clear characteristics for each category. They created a rubric and used it to score observations of a different object. This time, the task only took a half period, and students said, "Let's do it this way again." They bought into the concept of a rubric, because they experienced its power and efficiency.

Siciliano and Lach spent approximately two weeks introducing rubrics. They recommend that teachers spend ample time introducing rubrics. One field-test teacher commented about a module, "At first it was very difficult but it got much better as the students grew in their understanding." It should be noted that it took longer than two weeks for students to completely understand the process.

Once students understood the usefulness of basic rubrics, they were primed for more advanced rubrics. Lach and Siciliano introduced the SEPUP rubrics at this point, and showed students that they would be assessed on five variables (see Figure 1, p. 91) throughout the school year. The rubrics were placed over the walls of the classroom. One of Siciliano's students remarked, "Oh, to get a 3, I have to show that I did the experiment, I have the data, I've synthesized the data, and this is how it can be applied to my conclusion." The expectations were established and the focus became a process of learning to meet those expectations.

3. Customize the Rubrics

Because the SEPUP rubrics set the learning goals in the class, teachers must make sure that they match their own goals. SEPUP teachers sometimes added another rubric to the mix or adapted an existing rubric to better reflect their goals. Many teachers also recommended rewriting the rubrics in students' own words so students could see the connections between the rubric language and their own words. Teachers then posted the student rubrics around the classroom and referred to them. Donna Parker, a high school integrated science teacher in Columbus, Ohio, rewrote the Group Interaction Rubric using student language (Table 4).

Another SEPUP teacher adapted the Designing and Conducting Investigations Rubric to emphasize her goal regarding student predictions. The Designing Investigation element of this goal stated that a Level 2 response "[i]ncompletely states a problem or the design of an experiment." She adapted the criteria at each level to emphasize the importance of a prediction in the design of an experiment or investigation. Her Level 2 criteria were as follows: "Stated purpose is appropriate to the assignment, but may be incomplete. A prediction is mentioned, but not clearly stated or not related to the stated purpose."

4. Shift Students' Attention From Grades to Learning

In one classroom in Chicago, Doug Goodwin made comments to his students such as, "That looks like a 2 to me. How can you make it a 3?" His colleague, Marc Siciliano, wrote feedback questions that referred to getting a Level 4 for an answer and not settling for a 3. For instance, he stated that three pieces of evidence were needed in order to get a 4, even though students may have only learned two. Such expectations stretched students' thinking and encouraged connections to prior knowledge or extensions in order to achieve at high levels.

Goodwin and Siciliano constantly discussed terms like *almost there* and *complete and correct* with each other and their students, making sure that the message was consistent. In order for students to associate numbers with achievement, the teachers also named each level of the class rubric ("awesome," "cool," "embarrassing," and "bogus"). After all of the names were compiled, the class agreed on the following names for each level: "Above and Beyond" = 4; "Complete and

TABLE 4.
Customized Group Interaction (GI) Rubric in Student Language

Score	Task Management What to look for: Group stayed on task, managing time efficiently.	Group Participation What to look for: Group members worked together as a team and the ideas of all members were valued and weighed in working toward the common goal.
4 (A 50 pts)	Accomplishes Level 3 AND goes beyond in some significant way, e.g., group defines own approach to more effectively manage activity, group members actively help each other accomplish the task, group uses extra time productively. (Help other groups; assign roles; different people try things different ways; rework; extra work.)	Accomplishes Level 3 AND goes beyond in some significant way, e.g., group members actively <u>ask questions</u> about each other's ideas, group members <u>compromise</u> if there are disagreements, group members actively help <u>each other accomplish</u> the task.
3 (B 40 pts)	Group <u>managed time well</u> and <u>stayed</u> on task <u>throughout</u> the activity.	All group members participated and respectfully considered each others' ideas. (All working, contributing, listening. No arguing. No horseplay.)
2+ (C 35 pts) 2 (D 30 pts)	(Group works only when told.) Group stayed on task <u>most</u> of the time.	Unequal (one person does not do his or her part) group participation OR group respectfully considered some (only listen to people you like), but not all, ideas.
1 (F 20 pts) 0	Group was off-task majority of the time, but task completed. (Talking in groups or talking to other groups.) (Gossip. Not doing anything. Horseplay. Sleeping. Working on other subjects.) Group did not stay on task, which caused task not to be completed.	Significantly unequal group participation (one person does all work) OR group totally disregarded some members' comments and ideas. (Ignore group member. "Shut up." "That's stupid!" Talking while other members talk.) Single individual does entire task.
X	Group was not present.	

Note. Underlining was done by the teacher; parentheses and **different type** *indicate new text added based on student input.*

Source: Science Education for Public Understanding Program (SEPUP), Lawrence Hall of Science (University of California-Berkeley).

Correct" = 3; "Almost There" = 2; and "Major Errors" = 1. Developing the names helped students take ownership.

In these classrooms, students later questioned why certain of their responses had received a 2 rather than a 3. One student, examining an answer to an Evidence and Tradeoffs question, thought that the answer included sufficient relevant evidence. Siciliano explained that such dialogue demonstrated a greater level of student concern and understanding than did the students' questions during previous classes. This discussion opened the door for the teacher to focus on the additional evidence needed for a complete answer.

Rubrics also helped students view learning as a process. Siciliano's philosophy was that students start at different places on a continuum and some may take longer than others to reach Level 3 and 4 responses. He found something to build on in each student's work. Many of his students stated that rubrics motivated them in science. Having worked on a response only to have it marked with a big red X was much different than receiving a 2. There was no feedback from an X. A score, based on a rubric, clearly defined what was correct in the response and what could be improved.

5. Help Students Monitor Their Own Progress

In Chapter 12 of *Teaching Problems and the Problems of Teaching*, Lampert (2001) discussed ways to help students at different starting points improve in mathematics. One way is to foster independent learning by having students monitor their own progress (an aspect of metacognition). Students can do this by scoring their own and each other's work, thereby becoming more invested in the results. Another strategy, used by Chicago teacher Siciliano, is to have the teacher score student work and then simply tell students to check the rubric and figure out on their own why they received a particular score. This latter strategy demands practice and repetition, but gradually students do monitor their own learning. Siciliano found that, before he used rubrics, it took a lot of time to go over scored papers with students, because it was necessary to break down each problem. Rubrics help students understand what type of element is unclear in their responses, so they can be more in control of their learning.

An additional strategy is to have students keep a running record of their scores. Periodically, students reflect on their progress using their score records, and they make a note of any science questions or confusions they still have. This helps the student and teacher to monitor understanding and the teacher to adapt instruction.

6. Use Rubric Scores Within a Letter Grade System

Most school districts require letter grades and/or percentages for permanent records; however, this should not discourage teachers from using rubrics. Although Levels 0–4 may correspond with traditional grades, they should not be equated (a

4 should not be equated with an A, and a 2 should not be equated with a C). The rubric is a tool for documenting student development over time. Because Levels 3 and 4 of a SEPUP rubric describe major learning goals, it is unlikely that very many students would achieve a Level 3 at the beginning of the school year. Some students may be able to improve a scoring level over the course of a unit, while others will require a longer period of time to improve their performances. Thus, consistent Level 3s may not be necessary for a student to get an A or a B, especially at the beginning of a unit or course.

We encourage teachers to place rubrics into the contexts of their classes and local standards. The overall grading system is likely to include other criteria, such as completion of assignments and class participation. For example, out of a total of 20 points, a teacher may grade students' investigations and embedded assessments and assign a rubric score of 4, with the remaining 16 points determined by other criteria. Another teacher may decide that students who are able to improve by one scoring level (over a semester or a year) should receive an A or B for this aspect of work. A student who consistently scores 3s has demonstrated substantial competence, but no growth, and would not receive an A. Siciliano used spreadsheet formulas to convert rubric scores to more traditional grades for summative assessments. Adapting grading software also works well. Generally, these types of approaches satisfy the district, the parents, and those students who need to see a letter grade to confirm their successes.

Policy documents have increasingly referred to assessment practices as either summative (at the end of a unit or course to document learning) or formative (during instruction to inform learning). Using assessments formatively with students is an essential part of creating a culture of learning, although it does represent a major change for students and teachers (Bell and Cowie 2001).

CONCLUSION

Rubrics are tools for teachers and students. Teachers can clarify learning goals, give feedback, and help students build understanding through rubrics. Students can better understand learning goals, focus on learning rather than grades, and monitor their progress through rubrics. The development of an appropriate assessment system and classroom setup enables teachers to maximize the power of rubrics and use them to foster meaningful learning!

ACKNOWLEDGMENTS

We are grateful to our colleagues on the SEPUP and BEAR staff who contributed to development of the assessment system and to the following dedicated teachers for their input: Doug Goodwin, Michael Lach, Donna Parker, and Steve Torres.

REFERENCES

Bell, B., and B. Cowie. 2001. *Formative assessment and science education.* Dordrecht, The Netherlands: Kluwer Academic.

Lampert, M. 2001. *Teaching problems and the problems of teaching.* New Haven, CT: Yale University Press.

National Research Council (NRC). 2001. *Knowing what students know: The science and design of educational assessment.* Washington, DC: National Academy Press.

Roberts, L., and M. Wilson. 1998. *An integrated assessment system as a medium for teacher change and the organizational factors that mediate science teachers' professional development.* BEAR Report Series SA-98-2. Berkeley, CA: University of California.

Roberts, L., M. Wilson, and K. Draney. 1997. *The SEPUP Assessment System: An overview.* BEAR Report Series SA-97-1. Berkeley, CA: University of California.

Siegel, M. A., and M. A. Ranney. 2003. Developing the Changes in Attitude about the Relevance of Science (CARS) questionnaire and assessing two high school science classes. *Journal of Research in Science Teaching* 40 (8): 757–775.

Thier, H. D., and B. W. Nagle. 1994. Developing a model for issue-oriented science. In *STS education: International perspectives on reform,* eds. J. Soloman and G. Aikenhead. New York: Teachers College Press.

Wilson, M., and K. Sloane. 2000. From principles to practice: An embedded assessment system. *Applied Measurement in Education* 13 (2): 181–208.

Wilson, M., K. Sloane, L. Roberts, and R. Henke. 1995. *SEPUP Course I, Issues, Evidence and You: Achievement evidence from the pilot implementation.* BEAR Report Series SA-95-2. University of California, Berkeley (*www-gse.berkeley.edu/research/BEAR /publications/sepup95.pdf*).

RESOURCES

The Science Education for Public Understanding Program (SEPUP)

SEPUP courses include a student book with investigations and readings; an extensive teacher's guide that provides suggested teaching approaches, scientific background, information about students' ideas and possible responses, and suggestions for modifying or enhancing lessons for different student groups; and a full equipment kit. The modular materials provide student handouts as photocopy masters. More information is available about SEPUP courses and modules at *http://sepuplhs.org/currmat/index.html, http://sepuplhs. org/resources/assessment,* and *http://sepuplhs.org/profdev.*

One of SEPUP's professional development programs, Elementary Science Teacher Leadership (ESTL), has developed 10 guides for preservice and inservice teacher educators. Each guide includes activities, rationales, and resources to help teacher educators lead nine hours of professional development sessions. The Learning About Assessment guide is particularly relevant: *http://sepuplhs.org/profdev/index. html#teacher_educators.*

The National Center for Education Statistics (NCES)

NCES, in the U.S. Department of Education, is the primary federal resource for collecting and analyzing data related to education in the United States and other nations (*http://nces.ed.gov/practitioners/*).

The Center for the Assessment and Evaluation of Student Learning (CAESL)

CAESL seeks to improve student learning and understanding in science by focusing on effective assessment. The following link takes users to the resources page to find articles and video-based tools: *www.edgateway.net/cs/caesl/print/docs/310*.

Arter, J., and J. McTighe. 2001. *Scoring rubrics in the classroom: Using performance criteria for assessing and improving student performance.* Thousand Oaks, CA: Corwin.

This is a practical book that defines types of rubrics, provides help for developing new rubrics, and offers ways of using them to enhance learning. The resources section contains many sample rubrics to adopt or adapt.

Siegel, M. A., D. Markey, and S. Swann. 2005. Life science assessments for English learners: A teacher's resource. Unpublished manuscript. Berkeley, CA: University of California. Forthcoming at *www.lmri.ucsb.edu*.

This resource guide provides guidelines for assessments that are equitable and effective for English learners, assessments related to the National Science Education Standards for middle school life science, rubrics, teacher tips, and student work samples.

AUTHOR AFFILIATIONS

Marcelle A. Siegel studies and supports teacher and scientist partnerships at the Science and Health Education Partnership program at the University of California-San Francisco. She is principal investigator of "Investigating and Improving Science Learning and Assessment for Middle School Linguistic Minority Students," funded by the University of California Linguistic Minority Research Institute. She was a developer at the Science Education for Public Understanding Program (SEPUP) at the Lawrence Hall of Science, University of California at Berkeley, from 2001 to 2004 and a fellow at the National Institute for Science Education at the University of Wisconsin-Madison from 1999 to 2001.

Paul Hynds is a freshman biology teacher at Gilroy High School, in Gilroy, California, and was an instructional materials developer with SEPUP during 2000–2001. He has 15 years secondary teaching experience in science, 3 at the high school level, and 12 at the middle school level.

Marc Siciliano is the science director at the Young Women's Leadership Charter School in Chicago, Illinois. He directed the Math, Science, and Technology Academy (MSTA) for three years and taught science for five years at Lake View High School in Chicago. Currently a developer of instructional materials at Northwestern, he was a teacher adviser for SEPUP.

Barbara Nagle is director of SEPUP, and project coordinator for the development of SEPUP's two middle school courses, Issues, Evidence and You (IEY) and Science and Life Issues (SALI). She has a PhD in cell biology and taught high school chemistry in an inner-city Oakland school for six years before joining SEPUP as an instructional materials developer.

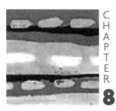

professional development and teacher learning: using concept maps in inquiry classrooms

Beth Kubitskey, Barry J. Fishman, Jon Margerum-Leys, Jay Fogleman, Stein Brunvand, and Ronald W. Marx

INTRODUCTION

In Ms. Gamble's middle school science class, the students finished their investigations of Newton's first law, seemingly engaged and actively learning complex material. James contemplated aloud, "What would happen if I had a ball on the handlebars of my bike and I stopped quickly," which prompted LaTasha to ask, "What would happen to the rider on the bike?"

Rather than giving a quiz at the end of the lesson, Ms. Gamble asked the students to develop concept maps to represent their understanding of the concepts they had studied. However, she was disappointed that the concept maps produced by students did not represent the learning she had observed during the activity. Ms. Gamble knew there would be other opportunities to assess her students' learning, but she pondered what she might have done differently to help students produce better concept maps. How could she adapt an alternative form of assessment to allow students to organize their thoughts about physics concepts? What could she do next time? Who could she call on to help?

This scenario led the authors to explore the role of professional development to support two teachers in implementing concept mapping as an alternative assessment. The teachers—Ms. Gamble and Ms. Case—are experienced middle school science teachers working in an urban setting and part of a large systemic reform effort.

NEW FORMS OF ASSESSMENT

Implementing authentic student assessment in the science classroom has proven challenging (Pellegrino, Chudowsky, and Glaser 2001). The intricate relationships among scientific concepts are complex, students and parents may not understand new types of assessments, and teachers need to learn how to integrate authentic assessment into the daily flow of classroom life. Teachers frequently attempt to discern what students know about science by testing their recall of facts or vocabulary. To really know how and what students learn from their classroom activities, teachers must use a variety of assessment tools. These can range from traditional assessments (e.g., tests, essays, and reports) to less conventional assessments (e.g., creation of artifacts, technology-based continuous assessment, and portfolios). Concept mapping is one form of assessment that affords teachers opportunities to observe and facilitate the connections that students make, both as a means of instruction and as an assessment tool (Novak 1990). Unless the less conventional kinds of assessment become commonplace in the classroom, they face problems with scalability and sustainability. To be scalable, an innovation needs to move from experimental use by a single teacher or a small group of teachers to mainstream use. Sustainability involves successful use of the idea over time and adaptation of the innovation to evolving objectives and changing support structure. Two ways in which teachers can learn innovative practices are by developing and implementing curriculum materials and by attending relevant professional development (Cohen and Hill 2001).

RESEARCH QUESTION, METHODS, AND ANALYSIS

We investigated the question "How does professional development influence teachers' use of concept mapping as a means of assessing student work?" To examine teachers' use of concept mapping as an assessment strategy, we analyzed two cases: middle school science teachers Sybil Gamble and Barb Case. Although similar methods of analysis were used, the data collection was slightly different in each case. The study of Ms. Gamble arose spontaneously through interactions between Ms. Gamble and the professional development team. The team consisted of the district science education coordinator and personnel from the University of Michigan (see Margerum-Leys, Fishman, and Peek-Brown 2004). Members of the research team recorded professional development activities. These field notes were collected through the use of a notes database (Emerson, Fretz, and Shaw 1995). The notes database was developed from tools used in previous research (Margerum-Leys 2002). Ms. Gamble voluntarily informed us about the impact of a professional development activity through an informal e-mail. The research team followed up by interviewing Ms. Gamble about the impact the activity had on her instruction and collected student artifacts.

The study of Ms. Case used data from three complementary sources: field notes from classroom observations; field notes from the professional development sessions; and interviews with Ms. Case and the other teachers who were part of the professional development. The research team conducted interviews with 10 middle school science teachers during the 2001–2002 school year using a semi-structured interview protocol. The flexible protocol enabled parallel cases to be constructed, while allowing interviewers to follow themes as they arose. The cases were selected based on how teachers described the influence of professional development activities on their classroom assessment strategies. Members of the research team developed and discussed analytic memos (based on their analyses of field notes and interviews); cases were written that illustrated these areas. This was an iterative process, with themes emerging from the data and influencing areas for exploration, which in turn influenced analysis of the data. Ms. Case was an example using this approach.

The concept-mapping activities that Ms. Case and Ms. Gamble used in their classrooms were part of an inquiry-based science curricula jointly developed by the Detroit Public Schools and the University of Michigan through their collaboration in the Center for Learning Technologies in Urban Schools (LeTUS). The units discussed here are two (of five) developed by LeTUS (Singer et al. 2000): "Why do I need to wear a helmet when I ride my bike?" (Schneider and the Center for Highly Interactive Computing in Education 2002) and "How can my good friends make me sick?" (Hug and the Center for Highly Interactive Computing in Education 2002).

Ms. Gamble was a science teacher in an urban middle school of 800 students, 99% African American, nearly all of whom qualified for a free or reduced-price lunch. Ms. Gamble participated in two LeTUS summer institutes and several monthly professional development (PD) sessions to support her implementation of LeTUS curricula. She has a strong science background. She taught high school biology for three years, was a nurse for 14 years, and then returned to teaching science at a middle school in 1999. She holds a BS in biology and environmental education and an MAT in science curriculum development.

Ms. Case taught at a middle school of 800 students in a working-class neighborhood. About half of the students were Latino, a third were African American, and the rest were white. Nearly all students were eligible for a free or reduced-price lunch. Ms. Case has a bachelor's degree in science education and is nearing completion of an MA in reading. She has been teaching middle school science for six years and has been an active member of the LeTUS teacher community since its inception in 1997.

Ms. Gamble implemented a LeTUS physics unit in which students learned about Newton's laws by investigating the driving question "Why do I need to wear a helmet when I ride my bike?" (Schneider and the Center for Highly Interactive Computing in Education 2002). One assessment strategy used in connection with an investigation of Newton's first law was to have the students draw concept maps showing their understanding of the scientific concepts (see Figure 1). Although students' maps varied considerably, they should have included basic concepts essential to the study of Newton's first law. The students brainstormed concepts as a group and then individually constructed maps. Later, students revisited their maps and developed group concept maps that were accompanied by a written description.

FIGURE 1.

Example of an Incomplete Student Concept Map Drawn at the Beginning of a Unit

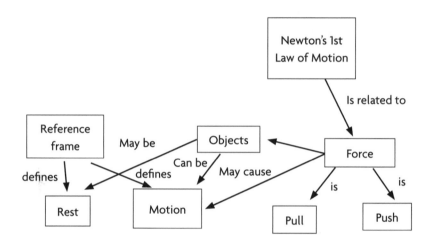

Ms. Case used a unit on communicable disease with the driving question "How can my good friends make me sick?" Early in the unit, students created concept maps of their initial understandings of the characteristics and transmission of communicable diseases. Following an investigation about the growth of bacteria, they revisited their initial maps to revise them or create new maps. A reproduction of a concept map created by Ms. Case's students during this unit is shown in Figure 2.

Two rubrics for evaluating the concept maps—a holistic and an analytical scoring rubric (Haney 1988)—were supplied in the curriculum and used in both units. The holistic rubric allowed teachers to rate the characteristics of the concept map using a 4-point scale, taking into account the number of branches, logical cross-links, completeness of concepts, and linking words. The analytical rubric allowed teachers to assess concepts, linking words, hierarchy, cross-links, examples, and form. The non-context-specific nature of the rubrics required both teachers to complete additional work to convert the rubrics from a general form to a more specific version suitable for assessing concept maps created by their own students.

Using concept maps as evidence of science learning was challenging. Through her interactions with her colleagues, Ms. Gamble shifted her assessment focus

FIGURE 2.

Concept Map Created by Ms. Cases's Middle School Students During a Unit on Communicable Diseases

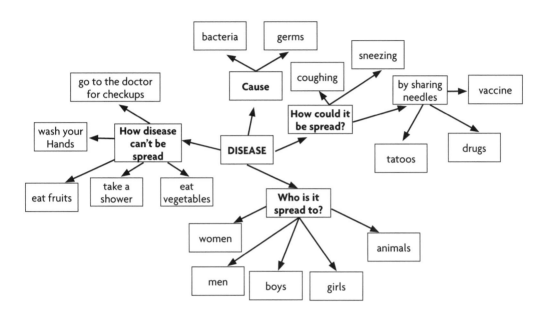

from assessing the form of the students' representations to interpreting the coherence of students' scientific understandings. Ms. Case used the students' concept maps to uncover student understanding through their discussions of the scientific ideas represented in the maps.

RESULTS

MS. GAMBLE'S STORY

A critical feature of our collaboration was a monthly, full-day professional development (PD) session at a partner school. These sessions featured teacher and curriculum developer presentations, discussions of student performance data and samples of student work, and time to share experiences about teaching the curriculum.

Ms. Gamble's experience demonstrated the benefit of her interactions with other teachers at these monthly PD sessions as she learned how to incorporate concept maps into her instruction and assessment. Before the PD sessions, Ms. Gamble's students had learned about Newton's first law through a cart-and-ramp investigation. They had brainstormed ideas and completed individual concept maps (following the lesson plan in the curriculum). Ms. Gamble had been disappointed because the students' concept maps were unsophisticated and lacked connections showing that they understood the topic.

Shortly after that experience, Ms. Gamble attended a monthly professional development session in which another teacher shared examples of her students' concept maps. The teachers evaluated the strengths and weaknesses of the examples. After a thorough discussion of the rubric in the curriculum and what each teacher thought was important for students to demonstrate, the teachers developed their own rubrics to evaluate the quality and completeness of students' work. The newly developed rubrics differed from the generic ones in the curriculum because of the content and assignment specificity. Using an iterative process, the teachers and the PD leader examined sample concept maps, created a context-specific rubric, and applied it to the samples. The rubric was made up of the following items:

* Driving question(s)
* Balanced and unbalanced forces (physics content knowledge)
* Descriptions or pictures describing previous activities and examples
* Vocabulary aligned with the school district's curriculum framework
* Statement of Newton's first law
* Pictures to depict key concepts
* Organizational links representing relationships between concepts

In addition to developing a new rubric, Ms. Gamble expressed to her colleagues in the PD sessions how important she felt it was to have had an opportunity to learn about what other teachers and students had done in the concept map activity. It gave her a better sense of expectations for her own students. She left the workshop inspired to repeat the assignment.

Ms. Gamble had students brainstorm their ideas and sketch concept maps. Next, students worked in groups to create group concept maps on poster paper. With a better understanding of what she might expect students to create and a clearer sense of how to assess the resulting maps, Ms. Gamble was able to give her students clearer directions for creating the maps. The resulting maps were more sophisticated and thoughtful, resembling the examples from the workshop, than the earlier maps had been. Ms. Gamble was pleased by these results, and she shared her experiences with the PD design team.

During interviews, Ms. Gamble expressed the importance of creating the rubric and her exposure to other teachers' ideas and student work as key features of professional development. She was not satisfied with her initial use of concept maps, and she stressed the problem of attempting to adopt a new curriculum for the first time. "There aren't a lot of assessment tools and I think [I] could get wrapped up in…the delivery of materials without stopping to think…how I was going to evaluate them." When asked what she found most beneficial about the workshop, Ms. Gamble again stressed the development of assessment tools: "The creation of the rubrics for the concept mapping…was for me the most beneficial…because it gave some serious and direct parameters for evaluating student work."

Professional development gave Ms. Gamble the opportunity to work with other teachers to examine the relationships and overall ideas reflected in the concept maps, rather than focusing on the specific details as she had previously done. She expressed this clearly in an e-mail message:

[I] have used concept maps before—but prior to workshops, more for specific content and/or detail. Since workshops, more for relationships and overall global concepts. Fine detail has become less important at the middle grades—if detail interests them, they will get it quickly, with or without me. If detail doesn't interest them, it remains meaningless to them. So my demanding little details at this level…is pointless. [I] don't believe it would help anyone perform on standardized tests, for that matter.

MS. CASE'S STORY

When informally assessing students, Ms. Case frequently asked questions to evaluate their understanding and encourage them to think scientifically. She described this way of working on concept maps with students:

I might ask, "Well, why did you put this [idea] with this [idea]?" and ask for an explanation of what they're putting down…. When they try to explain it and can't, that's when they know "Oh, that's not right," and then they'll change it…. So that's what my role is, facilitator and monitor.

Ms. Case's teaching style, which included anchoring questions in student work, informed her approach to assessing students' concept maps. While concept maps were valuable for assessment, they needed to be coupled with student explanations:

It's from the discussions…. If that discussion is taking place and they're explaining to each other why they want to put it down there or why it makes sense, then I know they're thinking in that way. That's really how I can tell. It's not a quiet thing, that you can just see them looking at the paper and know that they're thinking how you want them to. You have to have the discussion that goes along to let you know if they're getting it or not.

For Ms. Case, concept maps served to help students organize and demonstrate their understanding of science content. While concept maps were important assessment tools for her, they were also part of a more comprehensive strategy. This was an outgrowth of Ms. Case's effective teaching style (also demonstrated by her students' gains on LeTUS content evaluations).

Ms. Case's approach to assessment demonstrated an important point for curriculum developers: Teachers adapt the curriculum to fit to their teaching styles and context (McMillan-Culp and Honey 2000). Teachers integrate new strategies within their existing knowledge. Richardson (1997) noted that this idea goes beyond *prescribing* how learning *should* happen, to *describing* how it *does* happen.

Ms. Case valued the summer institutes and Saturday PD sessions as opportunities to talk with other teachers about effective teaching strategies. Opportunities for professional interactions with other teachers can be difficult to schedule; PD gave her the opportunity to make and continue connections with other teachers:

> *It's very helpful to me to talk to other teachers and hear what they're doing. We always are talking, because some of us are always at different places in the curriculum that we teach, and to hear from somebody "Well, I tried this and it worked really, really great, and kids got it" and they bring in stuff and say "Look what kids did" and it's wonderful. And then it's like "Oh, I never thought of that." And then you might try it or maybe even think "Maybe if I did this to change it just a little it'd be better."*

Since all teachers implemented the same units, there was common discussion of effective teaching methods during PD sessions. Having an understanding of the content and the context of the district, Ms. Case adapted other teachers' techniques to better fit her needs. One of the values of PD was the chance to advise colleagues about things that didn't turn out well:

> *You try something and, "What a bomb." It just does not work, it's way over their heads and totally [a] disaster. Why should someone else have to have a really terrible day if something's not going to go right, whether it's an experiment or just the whole management…of an activity?*

Ms. Case worked with LeTUS for six years, knew the university personnel and other teachers well, and was comfortable sharing her experiences, even when those experiences were not always positive.

DISCUSSION

ADAPTING CURRICULUM

When working with the curriculum units, teachers changed the units to fit their particular teaching styles, students' needs, and district goals. This adaptive process can be viewed as an aspect of teacher learning, in which knowledge gained from interacting with new materials was integrated into a larger body of knowledge. Adapting curriculum occurred with every teacher, to a greater or lesser extent, depending on particular situations.

Designers must explicitly address the need for adaptation of curriculum. They should anticipate such adaptation by suggesting alternative activities, providing prompts for teachers, identifying areas that are expressly intended to be modified by teachers, and building in opportunities for interaction among developers and teachers. Effective PD recognizes the need for adaptation by providing time for teachers to share their experiences and creating a community in which teachers feel comfortable. An important function of PD is the opportunity to help teachers learn how to take generalized assessment tools (e.g., rubrics for assessing curriculum) and customize them, making them more usable in their local contexts.

ASSESSING CONCEPT MAPS AS EVIDENCE OF SCIENCE LEARNING

Concept maps can provide assessment data that go beyond conventional measures, giving teachers better access to student understanding. However, our experience has shown that using concept maps effectively may require multiple attempts. Ms. Gamble was motivated to revise her use of concept maps and re-attempt an activity that was problematic in her first implementation because she was part of a community of science teachers. As Ms. Case's experience showed, concept maps were more effective assessments when they were coupled with oral or written explanations by students. Professional development helped Ms. Case to integrate concept mapping into her instruction and assessment system.

ACKNOWLEDGMENTS

We gratefully acknowledge the collaboration of Barbara Case and Sybil Gamble and their colleagues in the Detroit Public Schools. The research reported here was supported in part by the National Science Foundation (NSF) (REC 9876150 and 0830310A605) and the W. K. Kellogg Foundation. All opinions expressed in this work are the authors' and do not necessarily represent either of the funding agencies, Eastern Michigan University, or the University of Michigan. The research project was undertaken in the Center for Learning Technologies in Urban Schools (LeTUS), an NSF-funded effort that partners university researchers with teachers, administrators, and schools engaged in urban systemic reform. Researchers from the University of Michigan and Northwestern University collaborated with 75 educators from the Detroit and Chicago Public Schools to develop an agenda that promoted project-based science curricula with embedded technology in urban settings.

LINKS TO THE NATIONAL SCIENCE EDUCATION STANDARDS

The project described in this chapter has elements of professional development, curriculum design, and inquiry-based teaching and learning. In addition, our work with the Detroit Public Schools recognizes the value of diversity and the importance of high-quality science education for urban students. Ties to the National Science Education Standards (NRC 1996) can be made along each of these lines. Here are a few examples:

* Program Standard D acknowledges that "the most important resource is professional teachers." A key component of continuously improving teachers' skills and knowledge is a system program of professional development, as described in our chapter.

* Professional Development Standard A encourages "learning essential science content through the perspectives and methods of inquiry." The collaborative professional development effort between the University of Michigan and the Detroit Public Schools is delivered predominantly through inquiry learning.

* In this chapter, we identify *localizing curriculum* as an essential issue for curriculum designers. Both curriculum and the professional development that supports it must encourage and scaffold the cus-

tomization of curriculum. Program Standard D calls on teachers to "access to the world beyond the classroom." For each instructional setting, this access must be localizable for teachers and students.

❋ The Science Content Standards stress inquiry-based learning, extended time on task, and science as inquiry. All of these are hallmarks of HiCE, the program that sponsors the work described in our chapter.

❋ Program Standard E reminds us that "commitment to science for all implies inclusion of those who traditionally have not received encouragement and opportunity to pursue science." The urban setting in which we pursue our work is an environment in which African American and Latino students are provided with opportunities for excellence in science education.

REFERENCES

Cohen, D., and H. Hill. 2001. *Learning policy*. New Haven, CT: Yale University Press.

Emerson, R. M., R. I. Fretz, and L. L. Shaw. 1995. *Writing ethnographic fieldnotes*. Chicago: University of Chicago Press.

Haney, J. 1988. *Concept mapping in the science classroom: Linking theory to practice*. Agora, OH: Science Education Council of Ohio.

Hug, B., and the Center for Highly Interactive Computing in Education. 2002. How can my good friends make me sick? Curriculum materials. Ann Arbor, MI: University of Michigan.

Margerum-Leys, J., B. Fishman, and D. Peek-Brown. 2004. Lab partners: Research university and urban district join forces to promote standards-based student learning in science. *Journal of Staff Development* 25 (4): 38–42.

Margerum-Leys, J., and R. W. Marx. 2002. Teacher knowledge of educational technology: A case study of student/mentor teacher pairs. *Journal of Educational Computing Research* 26 (4): 427–462.

McMillan-Culp, K., and M. Honey. 2000. Scale and localization: The challenge of implementing what works. In *The Wingspread Conference on Technology's Role in Urban School Reform: Achieving Equity and Quality*, eds. M. Honey and C. Shookhoff, 41–46. Racine, WI: The Joyce Foundation, The Johnson Foundation, and the EDC Center for Children and Technology.

Novak, J. D. 1990. Concept mapping: A useful tool for science education. *Journal of Research in Science Teaching* 10: 923–949.

Pellegrino, J. W., N. Chudowsky, and R. Glaser, eds. 2001. *Knowing what students know: The science and design of educational assessment*. Washington, DC: National Academy Press.

Richardson, V., ed. 1997. *Constructivist teacher education: Building a world of new understandings*. London: Falmer Press.

Schneider, R., and the Center for Highly Interactive Computing in Education. 2002. Why do I need to wear a bike helmet? Curriculum materials. Ann Arbor, MI: University of Michigan.

Singer, J., R. W. Marx, J. Krajcik, and J. C. Chambers. 2000. Constructing extended inquiry projects: Curriculum materials for science education reform. *Educational Psychologist* 35: 165–178.

RESOURCES

LeTUS Curriculum

The curriculum units discussed in this article, developed by the Center for Highly Interactive Computing in Education to be used by LeTUS teachers, are described on the LeTUS website at *www.letus.org*.

Constructing and Assessing Concept Maps

Concept mapping is a way to assess students' understandings and refine instruction. Joe Novak presents an overview of concept mapping and evaluating concept maps on the web at *http://cmap.coginst.uwf.edu/info/*.

Strategies for using concept maps in classrooms can be found in J. S. Krajcik, C. M. Czerniak, and C. F. Berger. 2003. *Teaching children science elementary and middle school classrooms: A project based approach.* 2nd ed. Boston, MA: McGraw Hill.

Various software programs support classroom concept mapping. The most popular commercial programs are Inspiration (middle and high school) and Kidspiration (elementary school). Information about these programs is at *www.inspiration.com.* "The Concept Mapping Classroom," an article that describes how these programs have been used in classrooms, can be found at *www.education-world.com/a_tech/tech164.shtml.*

Discussing Student Work

Discussing student work with teacher colleagues can be an effective tool for refining instruction. Teachers have benefited from various approaches that support such discussions, including the Critical Friends Group initiative developed by the National School Reform Faculty. Information about training programs is available at *www.nsrfharmony.org.*

Deborah Bambino describes her experiences in a Critical Friends Group in the journal *Educational Leadership*; her article is listed in the ERIC database with accession number EJ-640974.

Sharon Cromwell describes how teacher groups lead to improvement in instruction in "Critical Friends Groups: Catalysts for School Change" at *www.education-world.com/a_admin/admin136.shtml.*

Lois Brown Easton discusses a "relatively risk-free way to present student work for feedback" in her *Journal of Staff Development* article "Tuning Protocols," available online at *www.nsdc.org/members/jsd/easton203.pdf.*

Kathleen Cushman recounts discussions about student artifacts in her article "Looking Collaboratively at Student Work: An Essential Toolkit" at *www.essentialschools.org/cs/resources/view/ces_res/57.*

AUTHOR AFFILIATIONS

Beth Kubitskey is a PhD candidate in teacher education at the University of Michigan, focusing her research on teacher learning from professional development. Her teaching experience includes eight years of teaching physics at Eastern Michigan University and four years of teaching high school physics, chemistry, and math.

Barry J. Fishman is an associate professor of learning technologies at the University of Michigan School of Education. His research focuses on the use of technology to support standards-based systemic school reform, models for teacher professional development in systemic reform, and the role of education leaders in fostering classroom-level reform involving technology.

Jon Margerum-Leys is an associate professor of educational technology in the Teacher Education Department at Eastern Michigan University. Margerum-Leys worked for seven years as a middle and high school teacher in New Hampshire and California. He was a Spencer Foundation research training fellow at the University of Michigan.

Jay Fogleman is a doctoral candidate and research assistant in science education at the University of Michigan. His background includes 11 years as a high school physics teacher in Maryland. His research focuses on the knowledge teachers need to enact classroom inquiry successfully.

Stein Brunvand is an assistant professor of educational technology at the University of Michigan, Dearborn. He taught elementary school for six years, both in Michigan and in the Marshall Islands.

Ronald W. Marx is a professor of educational psychology and the dean of education at the University of Arizona. His previous appointments were at Simon Fraser University and the University of Michigan, where he served as co-director of the Center for Highly Interactive Computing in Education and the Center for Learning Technologies in Urban Schools. His research focuses on how classrooms can be sites for learning that is highly motivated and cognitively engaging.

coming to see the invisible: assessing understanding in an inquiry-focused science classroom

Andrea M. Barton, Jennifer L. Cartier, and Angelo Collins

INTRODUCTION

For several years, science education researchers from the University of Wisconsin–Madison and science teachers at Monona Grove High School have maintained a collaborative relationship. One focus of our work together has been the design of curricula that mirror important aspects of realistic scientific practices. In particular, our curricula emphasize (a) the ways in which causal models are invoked to account for patterns in data and (b) the use of criteria when judging the acceptability of models and explanations.

Our own research (using field notes, interviews, and students' written work) has shown that students' participation in modeling and argumentation enables them to gain knowledge of Earth-Moon-Sun (EMS) concepts and improve their inquiry skills and understanding of how knowledge claims are judged in science (Cartier, Barton, and Mesmer 1999). Initially, students' progress in the unit was assessed based on data gathered almost exclusively from written work (mostly multiple-choice items assessing subject matter knowledge only). Using this kind of assessment data teachers did not "see" the strides students were making with respect to inquiry learning.

In this chapter, we describe a framework for thinking about components of assessment and discuss how we used this framework to create opportunities for teacher reflection about assessment decisions. Based on the teachers' reflections, we redesigned assessments for the EMS astronomy unit to help teachers systematically gather data on students' inquiry learning and focus on making students' inquiry skills "visible" through the use of data-gathering tools.

BACKGROUND

In the summer of 1998, a group of six high school science teachers and six university science education researchers (a group that has since come to be known as MUSE—Modeling for Understanding in Science Education) collaboratively developed a nine-week curriculum unit designed for a ninth-grade introductory science course. The unit began with two weeks of instruction that introduced students to science as a modeling endeavor through the use of a black box. During the following seven weeks, students developed a celestial motion model that they used to explain several near-Earth astronomy phenomena (e.g., sunrise and sunset, Moon phases, eclipses). Students learned how to

* collect and organize data,
* identify data patterns,
* develop and revise models, and
* share and assess explanations.

RESEARCH METHOD

MUSE focused primarily on curriculum development and student-learning research. All curriculum design work involved teachers in multiple capacities over three or more years. The research reported here represented initial efforts to employ a collective case study methodology to better understand how teachers adopted inquiry pedagogy and the types of professional development and support that had positive influences on these processes. The following data sources were used: daily field notes in a single Focus Classroom (taught by a teacher with eight years of experience); at least two days per week of field notes in classrooms taught by each of the three other Introductory Science teachers; daily audiotapes of the Focus Classroom; videotape of selected days of instruction in the Focus Classroom; notes (written and audiotaped) of curriculum development meetings between teachers and researchers; and curriculum materials, rubrics, and other written artifacts produced by the MUSE team during 1998–2000. Qualitative data such as interview transcripts and field notes were examined using QSR NUD*IST, a theory-building qualitative analysis tool.

FINDINGS

High school teachers and university researchers worked for over a year to revise a ninth-grade astronomy curriculum with the goal of incorporating more opportunities for students to learn through scientific practice. Contrary to recommendations in the National Science Education Standards (NRC 1996), however, assessments in this unit did not evolve along with the curriculum and instruction. Thus, when the new curriculum was first implemented, teachers used the existing assessments to measure student learning gains. These tools (mostly quizzes and exams) did not provide opportunities for students to demonstrate inquiry skills such as the ability to evaluate the consistency between evidence and claims or to apply knowledge to new situations. Instead, the assessment data documented students' factual knowledge about patterns (e.g., "How long does it take the Moon to orbit the Earth?") and elements of the Celestial Motion Model (e.g., "Which of the following describes why we have Moon phases?"). One of the distractors would be "The Moon orbits the Earth."

Because of their interactions with students in the class, teachers perceived that students were learning more than they demonstrated on these traditional assessments. Also, researchers interviewing students collected transcript evidence documenting that students were making gains in inquiry skills. Despite establishing both subject matter and inquiry-learning outcomes for students in the curriculum, it became apparent that only students' subject matter knowledge was being captured in the existing assessments. The team's next challenge became to make visible students' gains in inquiry skills through more authentic assessments that were better aligned with the unit's learning goals.

FIVE COMPONENTS OF ASSESSMENT

The MUSE group held biweekly, after-school meetings at Monona Grove High School throughout the 1998–1999 academic year. Angelo Collins, at that time a researcher at Vanderbilt University, joined a series of these meetings to facilitate thinking about the issue of assessment from a new perspective. She presented assessment as a collection of five related tasks:

* Identify learning outcomes.
* Design tasks that allow students to demonstrate mastery of those learning outcomes.
* Determine how you will gather the data related to those tasks.
* Develop a rubric that will help you make a judgment about the data with respect to a range of possible performances.
* Report about your judgment.

This perspective made a great deal of sense to the team. Rather than feeling overwhelmed by assessment as a whole, we came to see assessment as a set of manageable, interrelated components. After discussing these five components, the group agreed that assessment in the EMS unit had been breaking down at the "gathering data" stage. We then focused our attention to that particular component. The curriculum already included target learning outcomes and tasks that engaged students in aspects of authentic scientific practice. However, the curriculum did not include adequate tools for teachers to systematically gather data to help them make, and report on, accurate judgments about students' inquiry skills and ideas.

CURRICULUM REVISION

During the summer of 1999, we revised many of the data-gathering tools used in the EMS unit. We reflected on the entire EMS unit and created a matrix illustrating connections between each student activity and the learning outcome(s) addressed. This mapping activity helped us identify existing tasks and potential data that teachers could gather as evidence of students' inquiry learning.

THE BLACK BOX

The black box activity (Barton 2001; see also Cartier et al. 2005) introduced students to the processes of scientific modeling and argumentation. Student groups of three or four found a rather large box intriguing because of the phenomenon demonstrated (Figure 1). Each group poured water into their box and measured and recorded how much water flowed out. When a student poured in 400 ml of water, for example, sometimes no water, sometimes a small amount (about 200 ml), and sometimes a large amount (about 1,000 ml) of water flowed out. Over a two-week period, the boxes allowed students to experiment (gathering both quantitative and qualitative data and identifying data patterns) and develop, share, and assess models and explanations with their classmates and teacher. Students ultimately brought their models and data together by developing and sharing explanations that attempted to account for all available data, were consistent with the way they understood the world to work, and accurately predicted additional data patterns. The challenge was for *teachers* to gather their own data about students' inquiry learning from activities such as the black box.

FIGURE 1.
The Black Box Activity

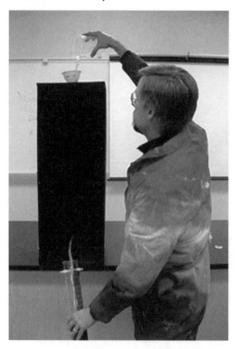

MAPPING THE FIVE ASSESSMENT COMPONENTS TO THE BLACK BOX ACTIVITY

To design assessment tools that teachers could use to systematically gather data and document gains in student inquiry skills, we were guided by Angelo Collins's framework (learning outcomes, tasks, data gathering, making a judgment, and providing a report). Before the development of such tools, teachers had collected bits of anecdotal evidence showing students making inquiry gains, but had no way to organize them when making judgments or reporting on student progress. The list on page 123 illustrates how we used the structure provided by the five assessment components to design a data-gathering tool called the presentation rubric (Figure 2). Teachers used this rubric to gather data during students' black box model presentations as well as to make judgments and report about students' performances.

FIGURE 2.

Presentation Rubric Developed by MUSE (Modeling for Understanding in Science Education) at the University of Wisconsin-Madison

Feature of the Performance	Rating			
	4	3	2	1
Technique	Speech able to be heard throughout the room Visual props show clarity and extra work Well-organized/practiced presentation Equal sharing of tasks among *all* group members obvious Presentation stayed within time limit No response to distractions by classmates	Speech able to be heard throughout the room Visual props able to be seen throughout the room Organized presentation Participation by all group members obvious Presentation stayed within time limit Minimal response to distractions by classmates	Speech able to be heard throughout the room Portions of the visual props are confusing Most of presentation organized Participation by all group members not obvious Presentation a little short Some response to distractions by classmates	Not everyone throughout the room was able to hear the speech Visual props absent or incomplete and confusing Unorganized presentation One or two group members were the focus of the presentation Presentation too short Easily distracted by classmates
Content	Clear introduction/problem statement Reasonable amount of data gathered Commentary *in addition to* a complete data table Offered insights into some procedural aspects Thorough summary Asked for audience questions	Clear introduction/problem statement Reasonable amount of data gathered Data table contains *all* data gathered Complete and clear explanation of procedures Thorough summary Asked for audience questions	Brief introduction/problem statement Reasonable amount of data gathered Data table contains most or all data gathered Complete but unclear explanation of procedures Brief summary Asked for audience questions	No introduction/problem statement Unreasonable amount of data gathered Little or no data shared in data table Sketchy/confusing explanation of procedures No summary Asked for audience questions
Argument	Strong match between model and supporting data Adequate discussion of model's conceptual consistency Clear evidence of model's ability to predict Discussed model's limitations/stated questions still to be pursued Able to answer most audience questions in depth	Solid match between model and supporting data Adequate discussion of model's conceptual consistency Clear evidence of model's ability to predict Stated some additional questions from the team Able to answer most audience questions	Parts of model's design supported by data Brief mention of model's conceptual consistency Limited evidence of model's ability to predict Few or no additional questions from the team Able to answer about 50% of audience questions	Model design is not supported by data No comment on model's conceptual consistency No demonstration of model's ability to predict Team has no additional questions of their own Unable to answer most audience questions

Learning outcomes. Students judged the acceptability of a model based on its explanatory power, its consistency with the way they understood the world to work, and its ability to predict empirical outcomes from new experiments (specifically, to predict how much water would come out of the box given a certain input volume).

Task. Students experimented with the black box, gathered data, identified patterns in their data, developed a model to account for their data, and shared that model with their classmates.

Gathering Data. Teachers used the presentation rubric to gather data about students' arguments on how well their models matched their data, were consistent with their understanding of the way the world worked, and were able to predict new data.

Judgment. At the top of the rubric was a range of performance levels (from 4 to 1) related to judging models based on three criteria: explanatory power, consistency, and predictive ability (see argument section of Figure 2).

Report. A simplified method of reporting involved mapping from rubric to grade such that 4 = A, 3 = B, 2 = C, 1 = D, and 0 = F. Ideally, a student's ability to judge models was determined based on his or her performance during several model-judging activities.

The target "learning outcome" and "task" listed above had been part of the black box activity before this new perspective on assessment was adopted. Teachers had not had assessment tools, however, to systematically gather data about how well students judged models during the black box model presentations. Before the unit was revised, student groups had presented their black box models to their classmates and teacher and designed poster representations; neither the presentation nor the poster had been judged. Teachers had listened to the presentations, questioned students about the ideas they presented, and collected students' posters. Without a data-gathering tool, teachers had not made judgments about the presentations or the posters.

Rubric design

We began with authentic student work (posters) produced in response to a task aligned with a specific inquiry-learning outcome (judging models). The team developed rubrics to help teachers systematically gather data from *both* student posters and presentations. Teachers would use their data to make a judgment and ultimately report on it. As professionals, the team members felt able to identify posters that represented exceptional (A), good (B), fair (C), acceptable (D), and unacceptable (F) levels of student work. Our challenge was to be explicit about, and come to consensus on, the criteria we used to make such judgments. At one after-school meeting, each of us chose five student posters to examine. We made judgments about our posters and then discussed our judgments as a group to identify the criteria we had used. Very interesting discussions occurred when two people held different judgments about the same poster. These differences suggested that people attached varying degrees of importance along three dimensions: technique, content, and argument. The poster and presentation rubrics described four levels of performance in each dimension and represented the consensus of this professional development experience. Despite beginning with student work samples and designing rubrics with performance levels informed by our best professional judgments, the rubrics needed to be refined to build teacher confidence in them as data-gathering tools.

Using the presentation rubric

Students began their EMS unit by experimenting with the black box. Their teacher and classmates listened to the group presentations and examined their posters. Rather than jotting down sketchy notes about each group's work, teachers used the presentation rubric to systematically gather data about student inquiry skills. One of the Monona Grove teachers commented on his use of the presentation rubric as a tool for systematic data gathering:

The rubric allows everyone to focus on the same set of elements. For example, in the argument section of the rubric, it is clear that an argument should do three things: explain all available data, predict additional data, and be consistent with other ideas about the way the world works. The teacher listens for students to make these points in a black box presentation and students know their argument should emphasize them…. The rubric provides a written record for students, teachers, and all others interested in a student's performance.

The same teacher made several other comments, however, that reflected his struggle to use the rubric with ease:

The rubric is cumbersome, specifically when attempting to gather data on the quality of a student's argument. For example, a student's argument may demonstrate a strong match between model and data (4 on the rubric), but provide no evidence of the model's predictive power (1 on the rubric). It is not uncommon for students to present an argument that is inconsistent in quality according to the elements of the rubric. This inconsistency makes it difficult for the teacher to make a judgment about the overall quality of a student's argument. Isolating each element in the dimensions of technique, content, and argument would help teachers gather more precise data, but would make the rubric more cumbersome to use as a tool. These are trade-offs inherent in performance-based assessment....

The rubric attempts to introduce a grade into the practice of science. Both students and scientists have their work and their ideas judged, but only students receive grades. The process of grading seems to take away some of the authenticity of students' experiences in aspects of scientific practice. No longer do they feel they are playing the game of science, but feel as though they are playing the game of school. From the students' perspective, introducing a grade into the activity makes it messy. This may be because students are not very skilled at presenting arguments during the black box unit that begins their ninth-grade science course. Thus they are more likely to score toward the 1 rather than the 4 on a number of elements. They are new to the practice of science and thus their initial grades are not very high. Not surprisingly, this is frustrating. There is not a given relationship between the rubric and a student's grade.

The rubric might be used to gather data during a student's presentation and then used again during a later presentation. These two sets of data could then be used as evidence of changes in a student's performance.

CONCLUSION

After three semesters of using the presentation rubric as a tool for systematic data gathering, teachers found it to be both helpful and in need of modification. The rubric represented an attempt to design a data-gathering tool to help teachers "see" inquiry skill gains students made as they participated in activities mirroring aspects of authentic scientific practices. The struggles that MUSE members have experienced in their attempts to design and use such tools are a testament to the challenges associated with engaging students in activities that resemble aspects of scientific practice and with assessing student performances on inquiry activities.

LINKS TO THE NATIONAL SCIENCE EDUCATION STANDARDS

The National Science Education Standards (NRC 1996) promote teaching and learning approaches that reflect authentic scientific practice, particularly the emphasis in science on inquiry as a way of achieving knowledge and understanding about the natural world.

As the Overview to the Standards states, "When engaging in inquiry, students describe objects and events, ask questions, construct explanations, test those explanations against current scientific knowledge, and communicate their ideas to others. They identify their assumptions, use critical and logical thinking, and consider alternative explanations. In this way, students actively develop their understanding of science by combining scientific knowledge with reasoning and thinking skills" (NRC 1996, p. 2).

If teaching and learning take an inquiry approach consistent with the Standards, it follows that assessment should keep pace and provide teachers (and others) with information about students' skills developed through inquiry activities. Through careful consideration of our instructional goals, tasks, information, judgment, and reporting, we attempted to match our assessment practices with the Standards.

REFERENCES

Barton, A. 2001. The Moon also rises: Investigating celestial motion models. *The Science Teacher* 68(6): 34–39.

Cartier, J., A. Barton, and K. Mesmer. 2001. Inquiry as a context for meaningful learning in a ninth-grade science unit. Paper presented at the National Association for Research in Science Teaching Conference, St. Louis, MO.

Cartier, J. L., C. M. Passmore, J. Stewart, and J. Willauer. 2005. Involving students in realistic scientific practice: Strategies for laying epistemological groundwork. In *Everyday matters in science and mathematics: Studies of complex classroom events*, eds. R. Nemirovsky, A. S. Rosebery, J. Solomon, and B. Warren, 267–298. Mahwah, NJ: Lawrence Erlbaum.

National Research Council (NRC). 1996. *National science education standards*. Washington DC: National Academy Press.

RESOURCES

The MUSE (Modeling for Understanding in Science education) website—*www.wcer.wisc.edu/ncisla/muse*—contains a complete collection of EMS (Earth-Moon-Sun) astronomy curriculum materials, a discussion of the MUSE philosophy on which the materials are based, and links to other relevant materials.

AUTHOR AFFILIATIONS

Andrea M. Barton teaches biology at LaFollette High School in Madison, Wisconsin, and is currently pursuing her PhD in science education at the University of Wisconsin (UW)-Madison. She has co-taught elementary science methods with Jennifer Cartier at UW-Madison and worked as a project assistant at the National Center for Improving Student Learning and Achievement in Mathematics and Science. She is studying the factors, including teacher professional development, that have an impact on the implementation of pedagogical approaches and curricula in a variety of school contexts.

Jennifer L. Cartier is an assistant professor of science education at the University of Pittsburgh, where she teaches science pedagogy (elementary and secondary) and curriculum theory/design courses. She received her PhD in curriculum and instruction (science education) from UW-Madison. Her research focuses on teacher professional development related to implementation of inquiry-focused instructional materials.

Angelo Collins taught high school biology for more than 15 years, has held various university appointments, and is currently the executive director of the Knowles Science Teaching Foundation. Her research interests lie at the intersection of science teaching, assessment, and policy. Her work on portfolios and other modes of alternative assessment informed the National Board for Professional Teaching Standards and contributed to the Interstate New Teacher Assessment and Support Consortium.

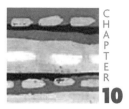

evolving ideas: assessment in an evolution course

Cynthia Passmore and Jim Stewart

A STORY ABOUT ASSESSMENT

On the first day of their nine-week elective evolution course, the seniors at Monona Grove High School in Monona, Wisconsin, completed a short writing assignment (see Figure 1). They wrote about how they thought a scientist would explain the origin of different shell shapes in two subspecies of Galapagos tortoises, those with saddleback and dome-shaped shells.

Their teacher, Ms. Johnson, had told them that this writing assignment would help her determine what they remembered from their sophomore biology class. She had emphasized that it would not be graded, but that it was important for her course planning that they thoughtfully complete the assignment. The students worked quietly on it, and handed their papers in at the end of the period. When Ms. Johnson looked over the responses she found that many students seemed to hold Lamarckian-like ideas, and that none of the students mentioned that the shells in the existing tortoise populations were variable. Moreover, their paragraphs were underdeveloped and did not convincingly bring evolutionary concepts together to explain change over time. Below are excerpts from some typical responses:

"They had to adapt to their surroundings by developing a shell that is higher up [saddleback]." (L.R.)

"In order for the Galapagos Turtles to survive they had to be able to reach the greens of the pear tree. In order to do that they need to stretch their necks and feet. They couldn't do that with their original dome-shaped shells so over time they adapted and their shells became like they are today [saddlebacked]." (S.E.)

"The saddleback carapace came into being due to the need of migrating tortoises to adjust to a new environment." (M.K.)

    Ms. Johnson used the responses, with help from the researchers who were working with her, to shape the instruction addressing many of the ideas she found in the tortoise explanations. For example, part of the course explicitly examined the ideas of Lamarck and compared them to those of Darwin. There were many opportunities for students to observe and represent variation in natural populations and to consider the role that variation plays in natural selection. In addition to learning about Darwin's natural selection model, the students had opportunities to use it to explain phenomena. By the end of the course, Ms. Johnson had evidence that the students made great strides in their understanding of evolution by natural selection, but she wanted to find out what the students thought they had learned.

    On one of the final days of the course, Ms. Johnson passed the original tortoise explanations back to the students and asked them to spend some time critiquing what they had written nine weeks earlier. As she explained this new assignment to them, several students groaned and complained that they "didn't know anything back then." Ms. Johnson was sympathetic and explained that she wanted them to be explicit about what they had learned and that one way to do that was to look back at their earlier ideas. As they had before, the students approached this assignment with care and were thorough in their critiques. Following are excerpts from what the students said:

"My whole explanation needs help. I talked about how the tortoises have to change to fit their environment. This statement is very much Lamarckian. Darwin believes mutations (variations) are naturally occurring. I needed to key in more about how the variation was advantageous...." (L.R.)

"...I also didn't recognize the trait's heritability, or the existence of variation in the past and present...." (S.E.)

"My original explanation makes the evolutionary process sound like a physical change taking place during the life of the tortoise and then being passed on to offspring. I now know that variations that are advantageous give animals a better chance of survival and allow them a better chance of passing on their advantageous trait to their offspring." (M.K.)

    In addition to critiquing their earlier explanations, Ms. Johnson asked students to write a new explanation for a different scenario. These two writing assignments, taken together with other assessments, provided compelling evidence that students not only learned the conceptual components of the natural selection model, but that they were able to use that knowledge to develop appropriately structured explanations and arguments.

FIGURE 1. ▨▨▨▨▨▨▨▨▨▨▨▨▨▨▨▨▨▨▨▨▨▨▨▨▨▨▨▨▨▨▨▨▨▨
Galapagos Tortoise Writing Prompt

The Galapagos Islands are home to giant tortoises that feed on green vegetation. Originally, tortoises lived only on Albemarle Island and had dome-shaped carapaces, or shells [original prompt included images of each type of tortoise]. Albemarle has a relatively wet climate and varied plant life.

Today tortoises are found on the other islands as well. On the small islands, such as Hood, there are tortoises that have a "saddleback" carapace. The "saddleback" carapace is elevated above the neck and flared above the hind feet. On the small islands the climate is drier and there is almost no ground vegetation. Prickly pear cactus (a major source of food and water for these tortoises) has a tree-like form; the woody trunk holds fleshy green parts of the plant high off the ground.

Using the information provided here, please answer the following question:

How might a scientist explain how tortoises with the "saddleback" carapace came to be? (Please give a detailed explanation and be sure to define any scientific terms you use.)

BACKGROUND

As science instruction moves from lectures that emphasize lists of facts to more student-centered approaches that emphasize knowledge generation and justification, it is clear that assessments must also change. The previous story is one example illustrating how students can engage in documenting and assessing their own learning.

At Monona Grove High School, teachers and university researchers wanted to develop a course that engaged students in activities mirroring the practices of evolutionary biology. Students could not simply describe the natural selection model; rather, the goal was to design a course that required students to use the model to explain phenomena (Passmore and Stewart 2002). This was a difficult task, but was followed with an even more daunting problem: What kinds of assessments would provide insight into students' abilities to perform like evolutionary biologists and, at the same time, provide feedback that could inform future iterations of the course?

RESEARCH STUDY, METHOD, AND ANALYSIS

The development and implementation of the evolution course was part of a larger research project that examined student learning and reasoning in model-based curricula. A scientific model is defined as a set of ideas about particular theoretical or real objects and the processes those objects undergo that can be used to explain natural phenomena (Cartier, Rudolph, and Stewart 2001). Our group, project MUSE (Modeling for Understanding in Science Education), developed curricula based on commitments to design instruction around key models in the discipline under study, take into account the ways in which scientists inquire within particular disciplines, and provide opportunities for students to develop, revise, and use models in ways that are true to the discipline.

The research study was conducted during an evolution course taught in spring 1999 in a medium-sized high school in the Midwest. The school drew students from a small city and from the surrounding, primarily rural, countryside. The evolution course was one of a set of nine-week science electives available to juniors and seniors. Students enrolled in the course were representative of the school population. Eleven females and seven males participated in the study.

In the evolution course, we organized the curriculum around one of the most important models in the discipline, Darwin's model of natural selection. (We have also developed curricula based on other important scientific models, such as meiosis, transmission genetics, and molecular models for a genetics unit, a model of Earth-Moon-Sun dynamics for an astronomy unit, and a model of kinetic molecular theory for a physical science course.) Based on the science education literature, we expected that students in our course would hold a variety of alternative conceptions about evolutionary mechanisms (Bishop and Anderson 1990; Brumby 1984; Halldén 1988). It was important to develop an assessment early in the course that would allow us to document some of the common ideas held by students. An additional consideration in the design of this assessment was that student work would be focused on the explanatory power of Darwin's model. Thus, we wanted to assess students' early abilities to develop explanations.

The study was designed to shed light on what students learned about Darwin's model of natural selection as a result of their participation in the evolution course. The data collected included pre- and posttests, students' written work during class, classroom field notes, student interviews, and audiotapes of student group and whole classroom discussions.

The initial analysis of students' ideas was conducted using written data sources. For the second part of the analysis, audiotapes were made on the days when students publicly defended their Darwinian explanations. The tapes were then transcribed; these transcripts were coded both for Darwinian concepts and for instances of argumentation and use of models.

The analysis of written work began with the final exam question. We developed a coding scheme that could be applied to both the pre- and posttest questions. A team of three researchers worked together to establish the codes, used them to analyze student work, and reconciled differences in coding. These meetings led to further clarifications in the coding scheme. The resulting codes focused on three areas: (1) how students described populations, (2) how students invoked a selective mechanism, and (3) how the language used in student explanations indicated their recognition of the probabilistic nature of natural selection. Once the procedure was solidified, two researchers coded responses together. Each researcher coded the work of all students in each area, and any differences in coding were discussed and resolved.

FINDINGS

The research conducted in the context of the evolution course documented the changes in student ideas over the duration of the course (Passmore, Stewart, and Mesmer 2001). The results indicated that, prior to instruction, many students held a number of alternate conceptions as reported in previous literature. Specifically, students did not invoke the natural selection model to explain change over time (in their early explanations, students focused on the present-day advantage of the saddleback shell).

Students' early explanations indicated that they did not understand the role of variation, reproduction, or heritability. Instead, they recalled common phrases such as "survival of the fittest," and used those kinds of terms without evidence of understanding them. Their explanations invoked Lamarckian mechanisms (such as need and use/disuse as the origin of traits) that would confer a survival advantage. They did not make explicit connections between the importance of survival *and* reproduction and heritability for the persistence of advantageous traits. The final writing assignments indicated that the students appreciated the connection between survival and reproduction and were able to discuss the importance of inheritance when writing Darwinian explanations. The natural selection model provided a framework for the explanation. A Darwinian explanation is a narrative that accounts for a change that has occurred in a population over time (Kitcher 1993). A complete explanation includes (1) a description of the trait at some time in the past, (2) reference to the variability in the population (this variability persists over time), (3) a description of the selective advantage of the trait and an explanation for how that advantage led to improved chances for survival and reproduction, (4) a discussion of the heritability of the trait, and (5) acknowledgment that the occurrence of the trait has increased in the population over many generations.

A key feature of the assessments in this model-based course was their fidelity to the intended outcomes. Our goal was to help students develop the capacity to reason like evolutionary biologists. The assessment tasks were designed for students to have repeated opportunities to do such reasoning and to receive feedback from their teacher and peers about their developing abilities to construct Darwinian explanations.

After several weeks of exploring and developing the natural selection model, students were given three case studies. Each case study required students to use the natural selection model in explaining some natural phenomenon. The first case study provided data and background information for a hypothetical plant species and geographic differences in seed coat characteristics among subpopulations. Students had to develop a poster that presented their arguments for how the subpopulations diverged. The development of the posters was an important feature of the instruction, and also served as an opportunity for the teacher (Ms. Johnson) to assess students' abilities in constructing Darwinian explanations.

The other two case studies required students to extend their understanding of the natural selection model to account for the seemingly counterintuitive instances of warning coloration, mimicry, and sexually dimorphic coloration. For the warning coloration and mimicry case, students worked several days on a paper about their Darwinian explanations, and they explicitly tied evidence from the case materials to their arguments. The sexual dimorphism case centered on a research grant competition, for which they developed a Darwinian explanation and a research proposal that would allow them to test some element of their explanations. The final products from the case work were an important part of the assessment program and accomplished two things. First, the final products gave students practice in model-based reasoning. In each case, students had to interpret the data and phenomena through the lens of the natural selection model. Second, the products provided an opportunity for students to receive formative feedback.

By the end of the course, students made significant gains in their understandings of variation, selection, and heritability. In both critiques of the Galapagos tortoise explanations and their new explanations, they showed increased sophistication in their use of concepts related to evolution by natural selection. The concept of selective advantage is a complex set of interrelated ideas. For students to grasp what it means for an organism to have an advantage in a Darwinian sense, they must recognize the interplay between survival, reproduction, and inheritance. At the beginning of the course, responses to the Galapagos tortoise writing prompt showed that many students associated natural selection with survival. What was missing was the recognition that simply surviving into old age was not sufficient in evolutionary terms. At the end of the course, students wove their understandings of individual concepts together to develop explanations for change over time in populations.

Although there is a great deal of value in providing opportunities for students to demonstrate learning in the ways described above, there are great challenges. These types of tasks are more time consuming to implement and grade than are more traditional forms of assessment. One way we alleviated this problem was to engage students in critiquing and evaluating one another's work. This was particularly useful in the formative stages. For instance, Ms. Johnson had students help develop the guidelines and expectations for a complete Darwinian explanation. These expectations were then used by students to judge the work of their peers. This technique eliminated some of the burden from the teacher in terms of time.

CONCLUSIONS AND IMPLICATIONS

Teachers have been using pretest and posttest assessments for years. The assessment instruments described here were important for instruction and research on students' science understanding. The writing assignment on the Galapagos tortoises provided insights for the teacher and researchers into students' early ideas. These insights shaped instruction and provided baseline data for analysis.

Throughout the course, students were asked repeatedly to develop explanations and were given feedback on their work. At the end of the course, the student critiques of their earlier explanations and their new explanations provided feedback on the efficacy of instruction. This information was used to shape instruction in subsequent years.

The initial assessment was used to guide instruction—Ms. Johnson used the insights she gained from the initial writing assignment to focus the subsequent lessons. Other teachers have found that this type of reflective teaching can enhance their students' learning (Treagust et al. 2001). The initial assessment was consistent with the goal of engaging students in realistic practice—students were asked on the first day of class to engage in evolutionary reasoning; they were provided with a task that called for an evolutionary explanation rather than isolated definitions of evolutionary concepts.

Developing the ability to construct explanations was an important goal of the course. The responses on the various writing assignments helped determine if that goal was met. All too often, there is a mismatch between curricular goals and what is valued on assessments. Students receive implicit messages about what is valued in the classroom based on the assessment tasks they are given. If the instruction and assessment do not match, learning often suffers (Bol and Strage 1996; Wright 2001).

An additional benefit of having students return to their own work is to help them appreciate how much they have learned. As one student stated after completing the second writing assignment:

> "Oh, man. Yeah. I think it is great. I knew the Darwinian Model coming in, the whole, you know, beneficial variation and stuff like that. But you just, then you get into the actual explanation part and that really, it really, you look at the world differently. I thought it was cool. Now I have this explanation in my head, and now I can run it through with whatever stuff I want. And I can try and develop a reason for why things are the way they are." (A.P.)

This student articulated the distinction between what he knew prior to the course and what he learned. Like many of his peers, he was familiar with the basic Darwinian model, but was unable to *use* that knowledge until he experienced tasks requiring him to construct explanations.

In summary, the assessments provided valuable information for the teacher, the students, and the researchers. Using the goals of the course, we designed a set of writing assignments that documented the evolving ideas of our students, and at the same time, allowed students to see how their ideas changed.

> **LINKS TO THE NATIONAL SCIENCE EDUCATION STANDARDS**
>
> The National Science Education Standards call for "authentic assessment exercises requir[ing] students to apply scientific knowledge and reasoning to situations…that approximate how scientists do their work" (NRC 1996, p. 78). The nine-week evolution course discussed in this chapter was designed to engage students in realistic inquiry that was true to the nature of evolutionary biology. Specifically, the course focused on the development of explanations: The students developed and critiqued evolutionary explanations for natural phenomena. These explanations served as both learning experiences and assessment exercises and helped students develop the "capacity to reason with knowledge" (NRC 1996, p. 91). Thus, the goals of our course and the assessment practices employed were consistent with recommendations found in the Standards.

REFERENCES

Bishop, B. A., and C. W. Anderson 1990. Student conceptions of natural selection and its role in evolution. *Journal of Research in Science Teaching* 27 (5): 415–427.

Bol, L., and A. Strage. 1996. The contradiction between teachers' instructional goals and their assessment practices in a high school biology courses. *Science Education* 80 (2): 145–163.

Brumby, M. 1984. Misconceptions about the concept of natural selection by medical biology students. *Science Education* 68: 493–503.

Cartier, J., J. Rudolph, and J. Stewart. 1999. The nature and structure of scientific models. NCISLMS Technical Paper. Available through the National Center for Improving Student Learning and Achievement in Mathematics and Science (*www.wcer.wisc.edu/ncisla/publications/reports/Models.pdf*).

Halldén, O. 1988. The evolution of the species: Pupil's perspectives and school perspectives. *International Journal of Science Education* 10: 541–552.

Kitcher, P. 1993. *The advancement of science: Science without legend, objectivity without illusions.* New York: Oxford University Press.

National Research Council (NRC). 1996. *National science education standards.* Washington, DC: National Academy Press.

Passmore, C., and J. Stewart. 2002. A modeling approach to teaching evolutionary biology in high schools. *Journal of Research in Science Teaching* 39 (3): 185–204.

Passmore, C. M., J. Stewart, and K. Mesmer. 2001. High school students' understanding of and reasoning with Darwin's natural selection model. Paper presented at the Annual Meeting of the National Association for Research in Science Teaching. St. Louis, MO (March).

Treagust, D. F., R. Jacobowitz, J. L. Gallagher, and J. Parker. 2001. Using assessment as a guide in teaching for understanding: A case study of a middle school science class learning about sound. *Science Education* 85: 137–157.

Wright, A. W. 2001. The ABCs of assessment. *The Science Teacher* 68 (7): 60–64.

RESOURCES

Mueller, J. The Authentic Assessment Toolbox. *http://jonathan.mueller.faculty.noctrl.edu/toolbox/index.htm.* This site contains "a how-to hypertext on creating authentic tasks, rubrics and standards for measuring and improving student learning."

Wiggins, G. 1990. The case for authentic assessment. *Practical Assessment, Research and Evaluation* 2 (2). Available online at *http://edresearch.org/pare/getvn.asp?v=2&n=2.*

AUTHOR AFFILIATIONS

Cynthia Passmore is an assistant professor in the School of Education at the University of California-Davis. She has collaborated with teachers to develop and implement innovative curricula in the areas of genetics, evolution, and astronomy. Her research focuses on student learning and problem solving in these areas with emphasis on long-term effects on student understanding. Prior to graduate work, she was a high school science teacher in California and Wisconsin.

Jim Stewart is a professor of science education at the University of Wisconsin (UW)-Madison. He teaches courses for prospective teachers and graduate courses on the nature of scientific inquiry, problem solving, and research design. He has extensive experience conducting institutes for science teachers, engaging in research on student learning and reasoning in the sciences, and collaborating with teachers on curriculum development. Prior to joining the UW-Madison faculty, Stewart was a high school biology teacher in upstate New York.

CHAPTER 11

varying instructional methods and assessment of students in high school chemistry

Avi Hofstein, Rachel Mamlok, and Otilia Rosenberg

INTRODUCTION

The release of the National Science Education Standards in the United States (NRC 1996) and in other countries around the world was a step of profound importance for the goal of achieving scientific literacy for all. The National Science Education Standards define the science content that all students should know and be able to understand. They also provide guidelines to assess the degree to which students have learned that content. The Standards include teaching strategies, professional development, and the support necessary to provide high-quality science education.

Because the content standards call for the implementation of new and varied pedagogical interventions and instructional techniques—requiring teachers to tailor instruction to meet the needs of diverse student populations—the assessment of students should also be multidimensional, drawing information from various sources. In recent years, similar reforms in science education were also implemented in Israel (Tomorrow 98 1992) and provide the context for the study described here.

APPROACH TO CHEMISTRY TEACHING AND LEARNING

In the last decade, science educators, including chemistry teachers and curriculum developers, have advocated that science in general, and chemistry in particular, should be taught not only to prepare students for academic careers in chemistry, but also to help them become scientifically literate citizens in a society that is highly affected by and aware of scientific advances and its technological manifestations (Bunce 1995; Kesner, Hofstein, and Ben-Zvi 1997). Consequently, chemistry should be taught with appropriate emphasis on its relevance to everyday life and its role in industry, technology, and society.

Recently, the chemistry curriculum changed dramatically—from focusing on a disciplinary approach to a multidimensional approach. This new approach included six dimensions (Kempa 1983): (1) the conceptual structure of chemistry, (2) the processes of chemistry, (3) the technological manifestation of chemistry, (4) chemistry as a personally relevant subject, (5) the cultural aspects of chemistry, and (6) the societal implications. This approach was highly aligned with the content standards that were already applied in chemistry curricula in the United States (ChemCom 1996), in the UK (Campbell et al. 1994), and in Israel (Hofstein and Mamlok 2001).

The introduction of these new dimensions into chemistry education was a call for a radical change in the teaching of chemistry, especially the use of different pedagogical standards tailored to the diverse cognitive abilities and aptitudes of students. The overall objective was to create a classroom learning environment that allowed students to interact physically and intellectually with appropriate instructional materials through hands-on, inquiry-oriented activities (Tobin, Capie, and Bettencourt 1988).

Hofstein and Walberg (1995) suggested that instructional techniques in science should be matched with the learners' characteristics, learning styles, and interests, in order to maximize the effectiveness of teaching and learning processes as well as to increase student motivation. Clearly, in practice, it is difficult to respond to each student's needs and learning style; however, much can be achieved if teachers employ a wide repertoire of instructional strategies. This enables many students to study chemistry in ways that are more aligned with their own interests and learning styles.

The implementation of a wide spectrum of instructional techniques and measurements of students' achievement and progress requires matching an assessment tool to each technique. According to the National Science Education Standards (NRC 1996), "Assessment policies and practices should be aligned with the goals, student expectations, and curriculum frameworks. Within the science program, the alignment of assessment with curriculum and teaching is one of the most critical pieces of science education reform" (p. 211).

In the past, only a small fraction of student learning activities in science were assessed (Lazarowitz and Tamir 1994). Students received final grades based mainly on their abilities on paper-and-pencil tests. Moreover, students' achievement when using other techniques (such as science inquiry laboratories, group and personal projects, reading scientific articles, and many other effective instructional activities) were rarely assessed and represented in students' final grades. This usually occurred because of a lack of valid and reliable criteria, as well as appropriate assessment tools. Another problem was that teachers were less confident about using alternative assessment methods.

RESEARCH QUESTIONS AND METHODS

We describe an experimental three-year project conducted in one high school in Israel, where chemistry was taught using new pedagogical and assessment standards. These school-based assessment procedures (planned and implemented by the chemistry teachers) replaced the centralized matriculation (Bagruth) examinations formulated by the Israeli government (these exams are part of the central education system and administered at the end of the 12th grade in Israel. The Bagruth measures only a limited scope of students' cognitive abilities and learning experiences [mainly using a one-time paper-and-pencil test].) The main idea behind this project was that the assessment should be valid, reliable, authentic (close to the students' way of learning), and comprehensive (i.e., it should cover a wide range of abilities and ensure meeting a wide range of learning goals).

The principal research questions were as follows:

* What are the students' attitudes toward the alternative assessment methods used to assess them?
* What are the teachers' attitudes toward the alternative assessment methods?

In order to probe into the students' attitudes and perceptions (about the project and the assessment method), the team of chemistry teachers developed a short questionnaire. This questionnaire was comprised of Likert-type items (using a 4 [fully agree] to 1 [totally disagree] scale). The questionnaire was administered to approximately 100 chemistry students at the end of the 12th grade (year three of the project). In addition, a series of interviews was conducted with several students to increase the construct validity of the questionnaire. Interviews were also conducted with the chemistry teachers.

The chemistry project occurred in an urban-type high school, with 100 students (in the same cohort over a period of three years) who studied chemistry between the 10th and 12th grades. These students opted to major in high school chemistry (10th-grade students studied chemistry for three periods per week and 11th- and 12th-grade students studied chemistry for eight periods per week).

PROFESSIONAL DEVELOPMENT, ASSESSMENT TOOLS, AND INSTRUCTION

A team of nine chemistry teachers was involved in the project, led by the school chemistry coordinator (head of department) and guided by an experienced tutor appointed to provide guidance for the project. The tutor, from the National Center for Chemistry Teachers at the Weizmann Institute of Science, was highly experienced in the development of learning materials, instructional techniques, and assessment of high school chemistry. The chemistry coordinator and tutor were involved as part of the team in the development of the instructional techniques and in the design of associated assessment tools. The assessment tools consisted of tests, quizzes, and assessment criteria for inquiry-type experiments, mini-projects, and critical readings of scientific papers. The teachers discussed and determined the weightings for each of the assessment criteria in order to increase objectivity of the assessments. The role of the tutor and the coordinator was to enhance the team's confidence regarding the importance and effectiveness of the project. They acted as guides for what could be regarded as long-term, dynamic, school-based professional development for the chemistry teachers. The team served as an "open platform" for dealing with problems that emerged from their daily work; they developed an environment of collegiality, where teachers reflected on their work and critically evaluated different pedagogical ideas and methods to improve students' assessment. This project could be regarded as action research, which nowadays is seen as central to the restructuring of schools and as a strategy for professional development (Loucks-Horsley et al. 1998). The school administration gave the team time and support to develop the program. The tutor provided access to research resources for teachers to develop the chemistry program. Finally, the team shared the results of the project with other chemistry teachers through teacher conferences held in various centers throughout the country.

LABORATORY-BASED INQUIRY EXPERIMENTS

Laboratory activities offer important insights to students regarding the learning of science in general, and chemistry in particular. These insights may be unavailable through other kinds of learning experiences (Lunetta 1998). For over a century, laboratory experiences have been used to promote desirable science education goals, including the enhancement of students' understanding of scientific concepts, scientific practical skills and problem-solving ability, and interest and motivation (Lazarowitz and Tamir 1994).

Learning by inquiry is an important component of the National Science Education Standards (NRC 1996). The inquiry laboratory can serve as an important vehicle for the development of many science skills, including planning and designing an investigation, conducting and recording observations, manipulating equipment, measuring and collecting data, analyzing and interpreting the data, asking questions about phenomena, suggesting hypotheses, and finally designing an experiment for further investigation. Instructions given to the students and the weights allocated to each of the procedural components of the inquiry-type experiments are shown in Table 1.

In a single lesson it was very difficult to compel students to conduct all components of the inquiry procedure. As the students gained more and more experience in conducting such experiments, they became more independent, and advanced their levels of inquiry.

MINI-PROJECTS

Mini-projects were activities involving a group of students aimed at in-depth understanding of a given question or issue in chemistry. The goal of the mini-project

TABLE 1.

Weightings and Instructions to Students for Components of Inquiry Experiments

Weighting	Instruction
Definition of the problem (10%)	Define the problem and make an assumption.
Planning the experiment (25%)	Plan your experiment accurately, logically, interestingly, and efficiently. Present your assumptions at each stage; act independently and prepare an equipment list.
Performance of the experiment (15%)	Follow the safety rules; use the proper tools and be careful with the materials.
Observation of phenomena (10%)	Carefully observe the materials and changes that occur during the experiment.
Analysis of data (30%)	Use concise and complete explanations; point out ambiguous observations, distinguish between assumptions, explanations, and conclusions.
Structure and format of the report (10%)	Ensure that your work is well-arranged, legible, and aesthetically presented.

was the development of greater independence and autonomy by the learner. Students obtained information from multiple resources (e.g., the library, the internet, and interviews with scientists). They defined a research question, discussed a research plan, and conducted an experiment in the school laboratory to explore their inquiries, using the results and data from the experiment for analysis in a written report. The mini-project assignment was carried out in small groups (four to five students in each group) and based on relevant issues. Most mini-projects focused on applications of chemistry to issues in the personal lives of the students. The projects incorporated a philosophy of science, technology, and society (STS) and were very much in agreement with several dimensions of chemistry education (Kempa 1983). Project topics included fluoride ions and the protection of the teeth, detergents in the household, vitamin C in citrus, the quality of drinking water, the chemistry of swimming pools, and the efficiency of various medicines in curing heartburn. Since the mini-project assignment included laboratory experiments, the instructions given to students were similar to those for the inquiry experiments (Table 2).

CRITICAL READINGS OF SCIENTIFIC ARTICLES IN NEWSPAPERS OR OTHER MEDIA
Scientific articles published in newspapers can be an important source for making subjects more relevant and up-to-date (Wellington 1991; Mamlok 1998). Critical reading of information from the media is an important part of developing science literacy.

TABLE 2.

Weightings and Instructions for Components of Mini-Projects

Weighting	Instruction
Research proposal (30%)	Define the objectives of your study, the research questions, and the assumptions; include a time planning chart and a literature review.
Performance of the experiment (15%)	Follow the safety rules; use the proper tools and be careful with the materials.
Results and discussion (30%)	Present the results; process the data; use concise and explicit explanations; define and distinguish between *assumptions, explanations,* and *conclusions.*
Structure and format of report (10%)	Ensure that your work is well arranged, legible, and aesthetically presented.
Presentation of the study (15%)	Present your project as clearly as possible (as a lecture, a poster, slides, or a PowerPoint presentation).

Each student chose one article from a collection of articles provided by the teachers. Article topics included important elements, the discovery of the rare Earth elements, chemistry in the Bible, thermodynamics and spontaneity, and chemical aspects of atmospheric pollution. The students were also given a written guide for critically reading the article (Table 3).

TABLE 3.

Weighting and Instructions for Critical Reading Components of Scientific Articles

Weighting	Instruction
Identification of concepts (15%)	Identify as many new concepts as possible.
Explanation of new concepts (20%)	Explain the new concepts. Use reference books and dictionaries. Note the source of each explanation.
Compilation of questions (40%)	The questions should be formulated clearly. The answers should appear in the article. Compile questions and raise criticism of the article's contents. Questions linking the article's contents with other fields studied in class may be used.
Answers to the questions (20%)	Use concise, complete answers.
Structure and format of the report (5%)	Ensure that your work is well-organized, legible, and aesthetically presented.

CONTINUOUS ASSESSMENT OF STUDENTS

Students' performance was assessed continuously as the students progressed from the 10th grade through the 12th grade. The weightings of percentages allocated for each of the assessment components assigned by the curriculum teaching team in grades 10, 11 and 12 are shown in Table 4. The weightings for each skill and assessment component were based on the discussions and decisions of the chemistry teachers' team.

TABLE 4.

Percentage Weightings for Student Assessment Components, Grades 10–12

Assessment Mode	Percent of Grade 10 (1997)	Percent of Grade 11 (1998)	Percent of Grade 12 (1999)
Examination	40	35	35
Quiz	10	10	15
Critical reading of scientific texts	20	20	20
Inquiry experiments	15	20	20
General performance[a]	15	15	10
TOTAL	100	100	100

[a]*General performance was based on the teacher's general impression of the student's attitude, dedication, and diligence.*

Students were aware of and involved with the assessment methods and their respective weightings. The teachers thought this increased students' responsibility for learning. The continuous assessment of students over three years gave students greater control over their achievements and progress in chemistry.

FINDINGS

The results from the questionnaire and the interviews (Table 5 and Table 6) indicated that, by and large, the participating students showed satisfaction with the new approach to assessments of their achievement and progress on various components of chemistry learning.

Interviews with the chemistry teachers revealed their satisfaction with the project, a significant change in their relationships with their fellow teachers, and a change in the general school climate. At the beginning of the project, few teachers had voiced their anxiety, feeling intimidated by the new approach; however, with the support of the group and cooperation from other team members and the tutor, the teachers gained greater confidence and became more flexible in their classrooms. They also stated that the new system of assessment was more time consuming than pen-and-pencil assessments, due to the following factors: the fixed team meetings, a lot of work with students and their assignments, and a heavy responsibility to assess students' portfolios.

TABLE 5.

Recommendations for Assessment Categories in Chemistry

Mini-Projects	Inquiry Experiments	Work on Scientific Articles	Quiz	Tests	Learning Objectives
✓	✓		✓	✓	Knowledge, comprehension, and application
✓	✓	✓		✓	Understanding of relationship between chemistry and the environment
✓		✓			Development of thinking skills in chemistry
✓	✓				Understanding of inquiry methods (and abilities) used in chemistry
		✓		✓	Understanding of the industrial applications in chemistry
		✓			Understanding of the nature and history of chemistry
✓	✓	✓			Development of scientific thinking
✓	✓				Development of practical (laboratory) abilities
✓					Using sources of information
✓	✓				Development of inquiry skills
✓		✓			Development of critical reading skills via scientific articles
✓		✓			Presentation of information, results, and data
✓					Development of ability to work individually
✓	✓				Development of ability to communicate scientifically with peers

Most teachers noted that the continuous assessments of students' progress and achievements provided a valid and reliable picture regarding students' knowledge and abilities. They also claimed that teaching this program changed their teaching habits. Teachers felt that they actually guided students in the learning process and did not merely deliver information to students. Student questionnaire responses and interviews corroborated the perceptions of their teachers. It should be noted that the positive feedback from students encouraged their teachers to continue with the project and to continue developing both their instruction and the aligned assessment methods.

DISCUSSION AND SUMMARY

In general, students who participated in this project expressed satisfaction with the way they learned chemistry. There was some evidence (based on the results of the action-research conducted by the teaching team) that the project established positive learning environments in the chemistry classrooms. These findings are especially important in an era in which creating an effective and improved classroom learning environment is a major goal. Moreover, there was no doubt that the new approach to assessment caused a significant decline in students' anxieties accompanying the matriculation examinations. Although students generally expressed satisfaction with the way they studied chemistry and were evaluated, they also perceived that they had to work harder than usual (compared with other school subjects) to obtain high grades. We also observed that there was a marked increase in the number of students who opted to major in chemistry. This again was an indicator of student satisfaction and the resulting publicity that the program received among the students.

TABLE 6.

Student Attitudes Toward the Assessment Method

Item	Mean Rating
The examinations and quizzes:	
Are well designed	3.63
Are returned on time	3.57
Allow me to demonstrate my knowledge	3.12
The teacher's comments are clear	3.29
Assessment of my achievement:	
It is clear to me how it is conducted.	3.38
The assessment is fair.	3.23
It is clear to me what I am suppose to do in the class.	3.42
My grades are related to my efforts.	2.56
Most of the lessons:	
Were interesting	3.05
Were clear	3.15
Provided challenge for further thoughts	3.05
The rate of the progress was adequate	3.45

Note: Responses ranged from 4-fully agree to 1-totally disagree.

In our opinion, this approach allowed for more intensive and effective interaction between students and learning materials, students and their peers, and students and their teachers. A summary of recommendations for two assessment tools, based on the project results, showed how assessments were aligned with learning objectives. We suggest that this approach could eventually increase students' motivation to learn and enhance learning, leading to increased science literacy.

LINKS TO THE NATIONAL SCIENCE EDUCATION STANDARDS

The National Science Education Standards (NRC 1996) state that "Assessment policies and practices should be aligned with the goals, student expectations, and curriculum frameworks. Within the science program, the alignment of assessment with curriculum and teaching is one of the most critical pieces of science education reform" (p. 211). The project described in this chapter focused on implementing new assessment standards as well as new teaching (pedagogical) standards. Students who studied chemistry in the project had ample opportunities to take control of their own learning by being involved in the development of the assessment tools used by their teachers. This project also applied the Teaching Standards (NRC 1996, p. 51). Teachers fully participated in planning and implementing professional growth and development strategies for themselves and their colleagues.

REFERENCES

Bunce, D.M. 1995. The quiet evolution in science education: Teaching students the way students learn. *Journal of College Science Teaching* 25: 169–171.

Campbell, B., R. Lazonby, P. Millar, P. Nicolson, P. Ramsden, and D. Waddington. 1994. Science: The Salters approach: A case study of the process of large-scale curriculum development. *Science Education* 78: 415–447.

ChemCom: Chemistry in the Community. 1996. Dubuque: Kendall/Hunt.

Hofstein, A., and R. Mamlok. 2001. From petroleum to tomatoes. *The Science Teacher* 68: 46–48.

Hofstein, A. and H. J. Walberg. 1995. Instructional strategies. In *Improving science education,* eds. B. J. Fraser and H. J. Walberg, 1–20. The National Society for the Study of Education.

Kempa, R. F. 1983. Developing new perspectives in chemistry education. In *Proceedings of the Seventh International Conference in Chemistry, Education, and Society,* eds. A. Rambaud. and H. Heikkinen, 34–42. Montpellier, France.

Kesner, M., A. Hofstein, and R. Ben-Zvi. 1997. The development and implementation of two industrial chemistry case studies for the Israeli high school chemistry curriculum. *International Journal of Science Education* 19 (5): 565–576.

Lazarowitz, R., and P. Tamir. 1998. Research on using laboratory instruction in science. In *Handbook of research in science teaching and learning,* ed. D. Gabel, 84–128. New York: Macmillan.

Loucks-Horsley, S., P. W. Hewson, N. Love, and K. Stiles. 1998. *Designing professional development for teachers of science and mathematics.* Thousand Oaks, CA: Corwin.

Lunetta, V. N. 1998. The school science laboratory: Historical perspectives and for contemporary teaching. In *International handbook of science education,* eds. B. Fraser and K. Tobin. Dodrecht: Kluwer Academic.

Mamlok, R. 1998. *Science: An ever-developing entity.* Rehovot, Israel: Weizmann Institute of Science.

National Research Council (NRC). 1996. *National science education standards.* Washington, DC: National Academy Press.

Tomorrow 98. 1992. *Report of the Superior Committee on Science Mathematics and Technology in Israel.* Jerusalem: Ministry of Education and Culture (English edition, 1994).

Tobin, K., W. Capie, and A. Bettencourt. 1988. Active teaching for higher cognitive learning in science. *International Journal of Science Education* 10: 17–27.

Wellington, J. 1991. Newspaper science, school science: Friends or enemies? *International Journal of Science Education* 13 (4): 363–372.

AUTHOR AFFILIATIONS

Avi Hofstein is a professor in the Department of Science Teaching at the Weizmann Institute of Science in Rehovot, Israel. Previously, he was a high school chemistry teacher for more than 12 years.

Rachel Mamlok is a staff scientist for the Department of Science Teaching at the Weizmann Institute of Science.

Otilia Rosenberg is a coordinating teacher of chemistry studies at Rishonim School, a large urban school in Hertzelia, Israel, and a regional tutor of chemistry. She is in charge of the alternative assessment of chemistry studies.

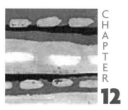

C
H
A
P
T
E
R
12

integrating an assessment plan into a k–12/university engineering partnership

Daniel W. Knight and Jacquelyn F. Sullivan

BACKGROUND

The goal of the Integrated Teaching and Learning (ITL) Program in the College of Engineering and Applied Science at the University of Colorado at Boulder was to provide a multidisciplinary learning environment that integrated engineering theory with practice and promoted creative, team-oriented problem-solving skills. Housed in the ITL Laboratory, a 34,400 square foot, technology-rich, hands-on learning facility, the program is equipped with design labs, measurement and analysis tools, team workrooms, manufacturing and electronics centers, and interactive learning exhibits (Carlson and Sullivan 2004). The ITL's K–12 Engineering Outreach Initiative extends the hands-on, inquiry-based learning vision into the K–12 community, using project-based engineering as a vehicle for the integration of math and science. It supports the belief that hands-on, active learning engages students in the excitement and satisfaction of gaining competency in science, technology, engineering, and math (STEM) (deGrazia et al. 2001; Schaefer, Sullivan, and Yowell 2003). Comprehensive assessment is also an important component of the initiative.

The outreach initiative targets three areas in its strategy to encourage young people to consider careers in engineering:

* The recruitment and preparation of university students serving as engineering "ambassadors" in Denver K–12 classrooms and after-school programs.
* The offering of summer engineering classes, workshops, and camps for K–12 students and teachers to enjoy the challenges of engineering.

* The development of standards-based K–12 engineering education curricula for use in K-12 classrooms across the country.

Supported by the National Science Foundation and the U.S. Department of Education, seven engineering graduate students serve yearlong fellowships as in-depth science and math content resources and engineer role models in K–12 classrooms. In collaboration with the K–12 teachers, the fellows introduce hands-on engineering activities to students as a real-world approach to theoretical science and math concepts (deGrazia et al. 2001).

During the summer, K–12 students and teachers participate in active-learning engineering classes and workshops that focus on specific STEM subject areas. For example, Creative Engineering was a weeklong electromechanical design-and-build class for students entering ninth grade. Mechanics Mania was a two-day, standards-based professional development workshop, in which upper-elementary-level teachers explored the importance of mechanics in solving everyday problems. In these workshops, engineering instructors and fellows from the University of Colorado at Boulder taught STEM fundamentals within an engineering context in fun and engaging ways (Poole, deGrazia, and Sullivan 2001; Schaefer et al. 2004).

Since the inception of the ITL K–12 Engineering Outreach Initiative in 1997, a robust collection of comprehensive engineering curricular units has been created for use in K–12 science and math classrooms. Topics include the laws of motion, airplanes, electricity, environment (air and water), energy of motion, and mechanics. Each lesson features affordable, hands-on activities designed especially for specific age ranges to strengthen youngsters' theoretical concepts and support their learning.

OUTREACH ASSESSMENT PLAN AND RESULTS

PROGRAM DEVELOPMENT

As the program was developed by ITL members, a detailed assessment plan provided structure for the curriculum. During this developmental phase, an assessment matrix was created to specify program goals, objectives, performance criteria, and assessment methods (Rogers and Sando 1996).

An assessment matrix specifying two of the seven goals that the fellows had for the teachers and students, as well as a portion of the objectives and criteria for each goal, is shown in Table 1. One goal is to improve teachers' skill and comfort with STEM content knowledge and pedagogy. This broad goal was translated into two more-specific learning objectives: (1) broaden an appreciation for inquiry-based, hands-on learning among teachers, and (2) avoid creating more work for the teachers. Criteria were developed to specify the precise level of performance required to meet each objective. Success was achieved if 90% of teachers indicated a willingness to participate in the program and if 90% of teachers reported that the program did not increase their workload.

The next step was to choose an assessment method to assess each criterion.

TABLE 1.

Excerpt of Assessment Matrix for Integrated Teaching and Learning (ITL) Program's K–12 Engineering Outreach Initiative

Goals	Objectives	Performance Criteria	Assessment Methods
1. Improve teachers' skill and comfort with STEM content knowledge and pedagogy.	1a. Broaden an appreciation for inquiry-based, hands-on learning among teachers.	1a. 90% of teachers indicate a willingness to participate again in the fellows-in-the-classroom program.	1a. Teacher semester survey.
	1b. Do not create more work for K–12 teachers.	1b. 90% of teachers report that the program did not increase their workload.	1b. Teacher semester survey.
2. Enrich STEM learning by K–12 students.	2a. Student learning is enhanced through experiencing relevant, hands-on, inquiry-based engineering curricula.	2a. Compared to control groups, treatment students demonstrate a significant increase in STEM content knowledge.	2a. Student content test.
	2b. Students gain knowledge of engineering as a career, an understanding of the pervasiveness of engineering in their world, and an appreciation of engineering as creating things for the benefit of society.	2b. At the end of the year, students demonstrate a greater understanding of engineering as a career.	2b. Student attitude survey, student focus group.

A survey was administered to all of the participating teachers at mid-year. Across each of the four years, teachers were asked if they wanted to continue with the program (yes or no) and to explain their answer. The results showed that 96% of teachers (109) wanted to continue with the program. One teacher stated, "I greatly appreciate giving my students exposure to 'real-life' application of science by working with fellows. Eighth graders benefit greatly by being able to ask questions and by having a connection with people pursuing science/engineering who aren't too much older than themselves." Information from this assessment tool indicated that program objectives were met.

PROGRAM IMPLEMENTATION

A comprehensive assessment plan is also important during the implementation phase to provide real-time formative feedback to program developers and instructors. A variety of formative assessment methods have been incorporated into the design of curricula for use in the K–12 classrooms, all targeted at answering the important question, "How do you know your students get it?" For example, one hands-on activity incorporated into a middle school environmental science lesson has students collect water and test for water quality. The activity is followed by a formative assessment activity in which students work in pairs to brainstorm other methods of testing water quality. The students then create a flowchart demonstrating their methods for collecting and treating drinking water. By examining the flowcharts, teachers gain insights into students' understanding of the concepts. Similar formative assessment activities can be found in Kagan (1994).

Formative assessment tools are used to provide K–12 students with feedback. A rubric of a formative assessment tool used by undergraduate mentors provided feedback to high school girls involved in a six-week summer internship called Girls Embrace Technology (Figure 1). The rubric is made up of measurement scales representative of the course objectives (Caso and Kenimer 2002). Rubrics typically have weighted anchors on each scale that provide an example of exceptional, fair, and poor performance (Davis et al. 1998). For example, the scale entitled "Participation" is anchored on the upper end by the following statement: "Participates actively and enthusiastically in the team project" and on the lower end by "Low level of participation; somewhat going-through-the-motions." The rubric, with its weighted anchors, is a useful rating tool for undergraduate team mentors who are generally inexperienced in giving performance ratings.

PROGRAM SUSTAINABILITY

Assessment also plays an important role in program sustainability, providing summative results for accountability. Summative assessment plans make use of qualitative and quantitative methods to assess four different targets: content knowledge, skills, affect, and opinions (Cunningham 1998; Payne 1994). Summative data are gathered and used at the end of outreach offerings to evaluate attainment of the goals and objectives.

The first summative assessment approach involves content testing. Changes in content knowledge are determined through the use of objective tests that ask students to choose from several answers (Cunningham 1998). Care was taken to ensure that course content was appropriately sampled. Up, Up and Away, a professional development workshop for teachers, explored the fundamentals of aerodynamics in a hands-on way. Twenty teachers anonymously completed a 10-item, multiple-choice test before and after the workshop. One sample item read as follows:

For a powered airplane in steady, level flight, drag:
   Is the sum of friction and pressure forces plus a part created from lift.
   Must balance the thrust.
   Must have the same magnitude as the lift.

FIGURE 1.

Formative Performance Feedback Rubric Used in a Summer High School Internship (Girls Embrace Technology)

Participation: Enthusiasm, involvement, and commitment

☐	☐	☐	☑	☐
Participates actively and enthusiastically in the team project.		Moderate level of participation; does what is necessary.	Low level of participation; somewhat going-through-the-motions.	

Information Technology Knowledge: Willingness and ability to learn new software

☐	☑	☐	☐	☐
Eager to learn new information technology knowledge; eager to take risks and stretch oneself to apply knowledge.		Somewhat willing to learn new information technology knowledge; makes some attempt to apply knowledge.	Unwilling to learn new information technology knowledge; makes little attempt to apply knowledge.	

Project Management Skills: Organization, Goal Setting, and Attainment

☐	☐	☐	☑	☐
Highly organized and efficient, pushes project toward goals; aware of time and deadlines.		Somewhat organized, works with structure provided by others; contributes to project goals.	Disorganized or has a tendency to procrastinate; distracts other team members from achieving project.	

The average pretest score was 44% across all items; a 75% posttest average demonstrated a 30% knowledge gain. T-test statistical procedures indicated a difference that was significant at the $p < .05$ level.

A second example of content testing was conducted with the fellows in the classroom program. The fellows, in cooperation with the K–12 teachers, developed content tests for 13 curricular units (i.e., Electricity and Magnetism, the Human Body, and Sound and Light). Fellows collected 1,139 matched pre- and posttests from eight schools (an experimental group in grades 3–12 and a matched control group that received a traditional curriculum) (see Figure 2). Analysis of variance statistical procedures revealed a significant interaction effect ($p < .01$), indicating that the experimental and control groups demonstrated different learning gains. While both test and control students began at the same place and both demonstrated significant gains, the ITL outreach group scored significantly higher than the control group on the posttest. In our opinion, summative data indicate that the ITL outreach curriculum and instruction more strongly affected student learning than did the traditional classroom curriculum and instruction.

We believe that this type of summative assessment had several advantages: Scoring could be accomplished with a high degree of accuracy (only one correct answer); the efficient format allows assessment to be inserted into a full curriculum; and external parties (such as granting and accreditation agencies) trust the results from objective tests. A strong criticism of this testing method is that it does not gauge the actual application of the knowledge (Cunningham 1998). Conceptual understanding is difficult to determine with a limited number of test questions.

Skills self-confidence is the second summative assessment tool used in ITL outreach initiatives. Students rate their confidence on a variety of skills associated with curriculum goals and objectives. Skills self-confidence is associated with motivation to pursue careers in engineering (Ponton et al. 2001). The 35 high school students attending Success Institute (a weeklong design-and-build camp targeting students typically underrepresented in engineering) reported their largest skills confidence gain as "giving technical presentations" (oral presentation experiences required in the curricula). Student confidence in oral presentation skills is developed through training in the use of PowerPoint software and rehearsing performances and conducting presentations to an audience of their family members, instructors, and peers. Student pre and post responses are a relatively efficient method of gaining insights without burdening instructors or training external judges.

A third summative assessment approach is the measurement of affective characteristics, such as interest in engineering and science, and comfort working in teams. Interest assessment predicts future behavior above and beyond the assessment of ability (Holland 1997). Similarly, comfort working in teams is an important predictor of success in engineering (Newport and Elms 1997). Student interest in majoring in engineering is measured at the beginning and end of the American Indian Upward Bound program (a six-week summer design-and-build initiative conducted at the University of Colorado at Boulder for recruiting Native American youth into the engineering field). Student interest in majoring in engineering jumped 33% in this program for the eight students, where students reported that they were "somewhat interested" at the beginning of the program and "highly interested" by the end of the program.

A fourth type of summative assessment is the use of the "opinionnaire," an open-ended assessment that ascertains student opinions about specific course components and program outcomes using broad, free-response questions (Payne 1994). Three open-ended questions are asked of participants at the conclusion of outreach classes and workshops: (1) What did you like best about the curriculum? (2) What did you find least satisfying about the curriculum? and (3) How could the curriculum be improved? One school principal associated with the fellows pro-

FIGURE 2.

K–12 Engineering Curricula Content Gains as Measured by Pre- and Posttests

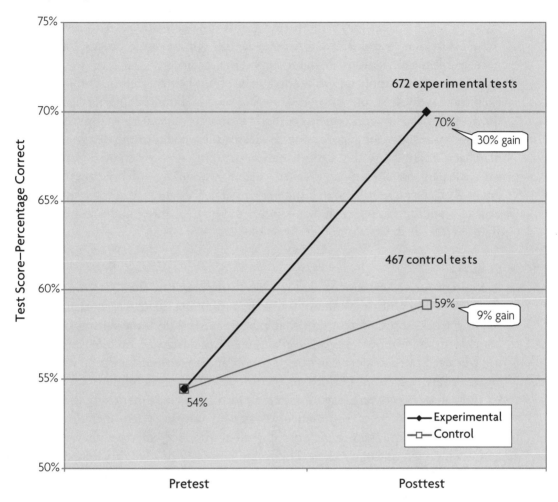

Note: Data were collected from eight schools, grades 3–12. The experimental group received the ITL curriculum and the control group received a traditional curriculum.

gram answered the opinionnaire question "What do you like best about the program?" by writing, "The quality of the fellows—they come in very knowledgeable and very eager to work with our students." Open-ended questions are a useful complement to quantitative data because the questions do not pre-forecast experiences by providing structured response categories and allow program participants to report information that may not have been considered by program planners.

LESSONS LEARNED

Two major lessons were learned during the implementation of the ITL Program's K–12 Engineering Outreach Assessment Plan.

Lesson 1: All parties must be aware of and committed to the assessment plan prior to program initiation. The development of an assessment matrix, securing informed consent, the administration of surveys, and looping assessment results into the program design are demanding and time-consuming. The rewards for these efforts are plentiful, but resistance to this cumbersome process can be expected unless expectations are agreed upon in advance.

Lesson 2: It is important to set up a debriefing shortly after the completion of each class, workshop, or camp. A debriefing session gathers all parties involved in the K–12/university components to review outcomes, share and discuss specific assessment results, and incorporate the lessons learned into the design of future plans and offerings. Without this timely concluding step, substantial time and resources spent on assessment may be wasted. Scheduling a debriefing places the analysis of assessment data on a shorter timeline and offers results for interpretation to a multidisciplinary audience with the goal of integration and informed changes (Poole, deGrazia, and Sullivan 2001).

SUMMARY

A comprehensive assessment plan is a key component of the ITL Program's K–12 Engineering Outreach Initiative. The plan provides an analytical structure for guiding workshop development, shaping implementation, measuring success at conclusion, and informing future planning. The design of our K–12 engineering outreach curricula typically begins with an assessment matrix that specifies course goals, objectives, performance criteria, and assessment methods. Curricular units incorporate an array of engaging formative assessment activities as well as the measurement of summative assessment targets. These include content-knowledge testing, assessment of skills self-confidence, affective characteristics, and open-ended questionnaires. The implementation results in valuable lessons learned. The importance of obtaining prior approval from the large number of parties involved in K–12 outreach and the importance of disseminating results in a timely fashion to assist future curricula design based on actual outcomes were significant findings.

LINKS TO THE NATIONAL SCIENCE EDUCATION STANDARDS

The work described in this chapter has links to two assessment standards of the National Science Education Standards (NRC 1996). The first is Assessment Standard A, assessments must be consistent with the decisions they are designed to inform. This standard suggests the need to deliberately design assessments in advance, which parallels the emphasis in this chapter on specifying an assessment matrix at the beginning of a programmatic effort. The second is Assessment Standard B, achievement and opportunity to learn science must be assessed. This standard is based on the need to assess the learning of content and the context in which the content is delivered. This process is discussed in this chapter in the section on summative assessment, as multiple assessment methods such as content testing and skills self-assessment are used for the purpose of determining program outcomes.

REFERENCES

Carlson, L. E., and J. F. Sullivan. 2004. Exploiting design to inspire interest in engineering across the K–16 curriculum. *International Journal of Engineering Education* 20: 372–378.

Caso, R., and A. Kenimer. 2002. Constructing rubrics for open-ended activities. Paper presented at the American Society of Engineering Education Annual Conference, Montreal, Canada.

Cunningham, G. K. 1998. *Assessment in the classroom*. Washington, DC: Falmer Press.

Davis, D. C., K. L. Gentili, D. E. Calkins, and M. S. Trevisan. 1998. *Program assessment of team-based engineering design: Concepts, methods, and materials*. Pullman, WA: Washington State University, Transferable Integrated Design Engineering Education (TIDEE) Project. Available at *www.tidee.cea.wsu.edu*.

deGrazia, J. L., J. F. Sullivan, L. E. Carlson, and D. W. Carlson. 2001. A K–12/university partnership: Creating tomorrow's engineers. *Journal of Engineering Education* 90: 557–563.

Holland, J. L. 1997. *Making vocational choices: A theory of vocational personalities and work environments*. Englewood Cliffs, NJ: Prentice Hall.

Kagan, S. 1994. *Cooperative learning*. San Juan Capistrano, CA: Kagan Cooperative Learning.

Newport, C. L., and D. G. Elms. 1997. Effective engineers. *International Journal of Engineering Education* 13: 325–332.

Payne, D. A. 1994. *Designing educational project and program evaluations*. Boston, MA: Kluwer Academic.

Ponton, M. K., J. H. Edmister, L. S. Ukeiley, and J. M. Seiner. 2001. Understanding the role of self-efficacy in engineering education. *Journal of Engineering Education* 90: 247–251.

Poole, S. J., J .L. deGrazia, and J. F. Sullivan. 2001. Assessing K–12 pre-engineering outreach programs. *Journal of Engineering Education* 90: 43-48.

Rogers, G. M. and J. K. Sando 1996. *Stepping ahead: An assessment plan development guide*. Terre Haute, IN: Rose-Hulman Institute of Technology, Office of Publications.

Schaefer, M. R., J. F. Sullivan, and J. L. Yowell 2003. Standards-based engineering curricula as a vehicle for K–12 science and math integration. Frontiers in Education Annual Conference, Boston, MA.

Schaefer, M. R., J. F. Sullivan, L. E. Carlson, and J. L. Yowell. 2004. Teachers teaching teachers: Linking K–12 engineering curricula with teacher professional development. Paper presented at the American Society for Engineering Education Annual Conference, Salt Lake City, UT.

Sullivan, J. F., M. N. Cyr, M. A. Mooney, R. F. Reitsma, N. C. Shaw, M. S. Zarske, and P. A. Klenk. 2005. The TeachEngineering Digital Library: Engineering Comes Alive for K–12 Youth. Paper delivered at the American Society for Engineering Education Annual Conference. Portland, OR.

RESOURCES

Chase, C. I. 1999. *Contemporary assessment for educators*. New York: Addison Wesley.

Erwin, T. H. 1991. *Assessing student learning and development*. San Francisco, CA: Jossey-Bass.

Gredler, M. E. 1998. *Classroom assessment and learning*. New York: Addison Wesley.

Knight, D. W., J. F. Sullivan, S. J. Poole, and L .E. Carlson. 2002. Skills assessment in hands-on learning. Paper delivered at the American Society for Engineering Education Annual Conference, Montreal, Canada.

Lissitz, R. W., and W. D. Schaffer. 2002. *Assessment in educational reform*. Boston: Allyn & Bacon.

TeachEngineering.com This online collection gives K–12 teachers and engineering faculty access to a rich resource of STEM (science, technology, engineering, and math) lessons and activities. Available at *www.cea.wsu.edu/TIDEE/monograph.html*.

AUTHOR AFFILIATIONS

Daniel W. Knight serves as the engineering assessment specialist in the Integrated Teaching and Learning (ITL) Program and Laboratory at the University of Colorado at Boulder. Knight provides team building and communications training for student participants in the ITL Program's outreach program and lower division engineering projects courses. He previously taught psychology after earning his doctorate in counseling psychology at the University of Tennessee.

Jacquelyn F. Sullivan is founding co-director of the Integrated Teaching and Learning (ITL) Program and Laboratory. She co-teaches Innovation and Invention and a service-learning Engineering Outreach Corps elective. Sullivan initiated the ITL's K–12 engineering outreach program. She has 14 years of engineering experience in industry and served for 9 years as the director of an interdisciplinary water resources decision support research center at the University of Colorado. She received her PhD in environmental health physics and toxicology from Purdue University.

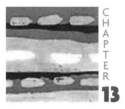

performance assessment tasks as a stimulus for collaboration among preservice and inservice teachers

Valarie L. Akerson, Amy Roth McDuffie, and Judith A. Morrison

INTRODUCTION

"We need to make sure that the penguins have the right environment."

"Yeah—they can't survive in Walla Walla's weather. We should put them kind of close to the polar bear exhibit so that the city planners realize that we took into consideration that the coolest areas were close to each other."

"Yes—good idea. I will write that note down so we can put it in the estimate letter to the planners."

Imagine the above conversation between third-grade students engaged in a science performance assessment task. Their goal was to design a zoo with at least five different climates, where each environment housed at least three different animals. They were then to design the environments for the animals and write a letter to city planners to convince them that this was a good plan for a zoo for their city. This actual performance assessment task was designed by two preservice elementary teachers, in partnership with an experienced inservice teacher. The task was implemented in the inservice teacher's classroom, allowing the teacher to assess her students' understandings of various animal habitats, design skills, and abilities to create maps.

BACKGROUND

National science education reform documents have recommended that students become more involved in their own learning, based on the philosophy that student understanding is facilitated by active involvement. These reform documents (e.g., AAAS 1993; NRC 1996) call for teaching that will motivate students to become reflective, constructive, and self-regulated learners. Education reforms require that students not only answer questions accurately, but be able to explain the process they used to derive their responses. The use of performance assessments can help with this reasoning process and is recommended to determine whether or not students can conceptualize important science concepts (Shymansky et al. 1997). Well-designed assessment tasks not only assess student understanding but also teach concepts (Darling-Hammond and Falk 1997; Shepard 2000). Performance assessment is particularly well suited to this purpose because it focuses on having students apply knowledge in an authentic context for an authentic purpose.

In science education, the National Science Education Standards (NRC 1996) state that the primary purpose of assessment should be to support the learning of important science. Assessment should be more than merely a test at the end of instruction to see how students perform under special conditions. To achieve this goal, the Standards called for embedding assessment in instruction, rather than keeping assessment as separate from learning.

Kelly and Kahle (1999) found that students who completed performance assessments were better able to explain their reasoning and conceptions than students who completed traditional tests. This finding led to the conclusion that they had stronger understandings, perhaps as a result of working through the performance assessment task. Shymansky et al. (1997) reported about the complications of designing performance assessment tasks that were truly valid.

Recognizing the value of performance assessment and the complexity of using these strategies, we (the authors of this chapter) decided to feature performance assessment in our science methods courses. This decision was part of our effort to prepare preservice teachers to use these approaches and to implement standards-based teaching and learning at the beginning of their careers. We decided on an established definition of performance assessment—Stenmark's (1991) statement that "performance assessment ... involves presenting students with a... task, project, or investigation, then observing, interviewing, and looking at their products to assess what they actually know and can do" (p. 13). In our performance assessment tasks, students were required to assume a role, establish a context and situation, and have an audience to which to present the final products.

Our decision to forge partnerships between preservice and inservice teachers to help inservice teachers experience and deepen their understanding of performance assessment was based on the literature on contextual and situated learning. Learning needs to occur in authentic activities in K–8 classrooms to improve the chance that what inservice teachers have gained from methods classes can be implemented in their own classroom instruction. For preservice teachers, a combination of university learning of theoretical foundations and school-based learning for a situated perspective is needed (Putnam and Borko 2000). Preservice and inservice teachers should participate in discourse communities as part of the learning and enculturation of the profession. Preservice teachers, in particular, need to learn about and contribute to a community's way of thinking (Putnam and Borko 2000). Spector (1999) recommended having preservice teachers work with inservice teachers to help them better apply newly learned teaching and assessment strategies. This finding is congruent with Dickinson et al.'s conclusion (1997) that important changes in science teaching can take place within the teaching context and with the support of an enthusiastic peer.

RESEARCH STUDY, METHODS, ANALYSIS, AND INTERVENTION

The goals of this project were to (a) help preservice teachers better understand performance assessment and increase their ability to design and implement science performance assessment tasks, and (b) increase the preservice teachers' classroom contact time with students and develop a partnership with inservice teachers.

The current study occurred during a one-semester K–8 science methods course. Nineteen students, working toward a master's in teaching degree, were enrolled in this science methods course (the only course they would complete to prepare them to teach science). In addition to designing and administering the performance assessment task with the help of their mentor inservice teacher, other assignments were to (a) study a science content area, then design and administer an interview of a K–8 student to identify student misconceptions, (b) design lessons to address student misconceptions, (c) participate in hands-on, minds-on activities in their science methods class, (d) submit weekly reflection papers on assigned topics, and (e) participate in weekly activities designed to improve nature of science conceptions.

DATA COLLECTION AND ANALYSIS

Prior to the introduction of the project, a baseline understanding of the preservice teachers' conceptions of performance assessment was collected through surveys and interviews of all 10 students who met the interview criteria. (These students were concurrently enrolled in an advanced educational psychology course and taking math methods in the spring. These criteria were used to ensure that the selected group of preservice teachers were part of the same cohort.) Both researchers kept independent logs of their perceptions of preservice teacher understandings and the challenges of program implementation. Student reflections were collected, and videotapes were made of several preservice teachers administering their tasks. (See Appendix A on p. 168 for interview questions.)

To address logistical issues in the field-based component, the researcher logs, discussions and e-mails with preservice teachers, and final performance assessment task reports were collected and analyzed. These data allowed us to develop a profile of the project in order to make recommendations for future implementations.

Data were collected from inservice teachers through a survey (see Appendix B, p. 169) mailed after the preservice teachers completed their field work. The inservice teachers were asked about their understandings of performance assessment, how they implement performance assessments in their classrooms, and their ratings of the preservice teacher's performance assessment task implementations. The inservice teachers were also asked if they learned anything about performance assessment through their mentorship experiences and if they had suggestions for future projects.

The researchers and a graduate student sorted through all data. These data were analyzed to develop early categories in response to the research questions. Preservice teacher responses to interview questions were coded using the scheme developed by Fuchs et al. (1999). Interview responses were coded as either a 1 or a 0, indicating whether preservice teachers included items that showed understanding of performance assessment. The final performance assessment tasks completed by the preservice teachers in collaboration with the inservice teachers were coded with the same scheme. It should be noted that the coding scheme did not determine how well the tasks were developed, but whether the responses or tasks included components indicating how well teachers understood performance assessment.

INTERVENTION

A collaborative team composed of a science educator (Akerson), a mathematics educator (McDuffie), two middle school mathematics teachers (Droppo and Fulton), a math/science coordinator from the Educational Service District (Morrison), and a secondary program administrator from a Washington State Educational Service District (Kirby) planned the performance assessment project at the beginning of the semester and adjusted the project as needed. (Ms. Droppo and Mr. Fulton were regionally recognized teacher-leaders for their expertise in performance assessment strategies and implementing standards-based approaches to teaching and learning. During the semester, they recommended other teachers for the cadre of mentor teachers.) A key component of the program was providing meaningful interactions between preservice teachers and inservice teachers to facilitate the preservice teachers' enculturation to the teaching community (Putnam and Borko 2000).

Introductory assessment workshop

A three-hour workshop that paired the preservice and inservice teachers was held during the science methods course to enable the teachers to work together on the design of the assessment tasks with the support of the collaborative team. The workshop was conducted during the regular methods class meeting time. The collaborative team planned and facilitated the workshop, with team members leading different parts of the workshop. The goal was to (a) briefly discuss general assessment issues, (b) provide an overview of the standards-based assessment program in Washington State (e.g., Washington Commission on Student Learning

1998), and (c) introduce preservice teachers to performance assessment issues and strategies.

To introduce the preservice teachers to performance assessment, we asked them to work in groups on a sample performance assessment task that was written and field-tested as part of an assessment program. The task required the preservice teachers to design a cereal box that would reduce the amount of cardboard needed but still maintain a specific volume, and then to write a letter to the cereal company describing and defending their design. They had sufficient time to identify key issues of the task and key components of task-design. Next, we discussed some features of the task (e.g., an open-ended question; the descriptive and persuasive writing component; the multiple entry points and various solution methods possible in performing the task). They were able to begin to design their own boxes on paper, but in the interest of time we then gave the groups scoring rubrics and samples of ninth-grade students' work, including sample written work and completed boxes that illustrated various performance levels. Using the scoring rubrics, the groups assigned scores to their sample student work. Following this, we discussed the scoring process, the rubrics, and the task as a class.

We worked to formalize their knowledge of performance assessment by discussing defining characteristics—the advantages and limitations—of performance assessment. A middle grades language arts teacher-leader facilitated a brief discussion of the types of writing used in performance assessment (e.g., descriptive, expository, persuasive). We concluded the workshop with an introduction of the planning guide (see Figure 1) and provided time for generating ideas for the preservice teachers' Performance Assessment (PA) projects while the inservice teachers were available to answer questions.

Researching topics and generating a plan for the PA task

The preservice teachers worked individually to generate performance task ideas related to their content areas. The PA task was linked to their earlier student interviews and lesson plan assignments. The preservice teachers selected a range of grade levels at which they would be most interested in designing and implementing a PA task.

The preservice teachers submitted planning guides (Figure 1) outlining the major features of their tasks. An important part of this planning guide was aligning the task with standards for learning. Because the Washington State Essential Academic Learning Requirements (EALRs) (Washington Commission on Student Learning 1998) were emphasized in this course, our students identified appropriate EALRs for their tasks. The groups continued developing their tasks outside class. While many groups created original tasks, they were permitted to use outside resources (e.g., activity books, journal articles) for ideas. Even in cases where a problem, activity, or task was used from an outside source, significant work was required to develop the problem into a PA task and meet the assignment requirements.

FIGURE 1.
Planning Guide for Developing a Performance Assessment Task

Title of Task:

Grade Level(s):

Concepts & Processes Assessed:

Task Context/Situation:

Students' Role in Performing the Task:

Audience/Client/Customers for whom the task product and/or performance is being created:

EALR* Benchmark (or NSES*) Assessed:	Student Performances/ Products:	Criteria for Evaluating Products/Performances:

Note: This planning guide was used by preservice teachers to generate ideas in the required categories for performance assessment development. The preservice teachers received feedback from the science methods instructor and their inservice mentors on their guide contents.

*EALR = Washington State's Essential Academic Learning Requirements; NSES = National Science Education Standards

Source: Adapted from McTigue, J. 1999. Performance task blueprint. In Second performance assessment summer workshop manual, ed. Washington State Office of Superintendent of Public Instruction. Seattle, WA: Office of Superintendent of Public Instruction.

Matching mentors and preservice teachers

Using the information provided in the preservice teachers' planning guides, we matched each preservice teacher to a mentor inservice teacher, based on the topic, skills, abilities, and level of thinking required for the PA task and the knowledge and grade level of the mentor teachers' students. The preservice-inservice teams initially met on their own after contacting one another by phone or e-mail. Mentors were sometimes assigned to more than one preservice teacher.

The mentors attended the methods class. Preservice teachers brought their planning guides and drafts of their PA tasks to class and met with mentor teachers to discuss their PA tasks. Additionally, other members of the planning team were available to assist groups in designing their tasks.

Submitting the first draft and field-testing the PA task

During the eighth week of the semester, the groups submitted their first drafts to their science methods professor and mentor teacher. Within a week, the science methods professor and inservice mentor teacher provided written feedback and comments for the groups to consider before administering their tasks to students.

Each group arranged a time to field-test their PA tasks in their mentors' classrooms. The tasks were designed to be completed in one to three 50-minute class periods. Each mentor teacher decided with his or her groups who would facilitate the tasks. In some cases, the mentor teacher was the primary facilitator, and in other cases, preservice teachers facilitated the task. In all cases, preservice teachers observed task administration, talked with students (in some cases, interviewed students about their thinking), and recorded notes on the process.

Following the field-test, preservice teachers scored the students' work and analyzed selected students' work in greater depth. They prepared a written report of their findings and reflections on the performance assessment process and projects.

RESULTS

UNDERSTANDINGS OF PERFORMANCE ASSESSMENT

Prior to the project, preservice teachers exhibited poor understanding of performance assessment, indicated by low scores on the coding scheme (see Appendix C, p. 170) (Fuchs et al. 1999). Preservice teachers included very few components necessary for a performance assessment task; their examples tended to be short, required single answers, and did not provide opportunities for their students to generate ideas. None of the preservice teachers required students to explain their work or to write about their work. Their ideas of performance assessment were not couched in authentic tasks.

Following the intervention, especially after developing and administering their own tasks, preservice teachers' understandings of performance assessment improved (indicated by scores on the coding scheme). All preservice teachers required written explanation of strategies, modeling of strategies, and multiple questions requiring application of knowledge set in an authentic context from their students. Most tasks required students to generate ideas and information rather than memorize or provide single-answer responses. For example, one student designed a performance assessment task entitled "To Market, to Market" in which fourth-grade students designed a method for transporting goods to market for a Kingdom (context for task). The students were members of the Royal Engineering Group (the role), needed to think about concepts they learned earlier about friction to apply in their design, and then write a letter to the king and top advisers (audience) to convince them that their design was best in terms of science content, materials, and cost. (See Table 1 for a list of the performance tasks designed by the groups.)

PERCEPTIONS OF MENTOR TEACHERS

Mentor inservice teachers provided feedback through a survey. Some of the mentor teachers had completed an introduction to performance assessment (usually through workshops provided in their schools) and others had no experience with performance assessment. They did not fully understand performance assessment and could not adequately rate the preservice teacher's implementation of the PA tasks. The inservice teachers also improved their *own* conceptions of performance assessment by working with the preservice teachers. Feedback on the project was positive from the mentor teachers, where one teacher stated that she thought it was "another excellent way for students to get into the classroom."

PERCEPTIONS FROM THE EDUCATIONAL SERVICE DISTRICT (ESD)

The collaboration between the ESD and the university allowed this project to begin building strong relationships among university faculty and students, inservice teachers, and ESD employees. The ESD offered stipends and release time to participating inservice mentor teachers to attend the methods course workshops. Mentor teachers involved in the project had little experience with or knowledge about performance assessment. As a result of the project, the ESD committed to involve more inservice teachers in performance assessment training workshops and experiences.

DISCUSSION AND IMPLICATIONS

The most positive comments from preservice teachers were that they were pleased to be working with actual students. This finding was consistent with Putnam and Borko (2000), that is, that a situated perspective is most meaningful to learning. When asked whether they would rather teach a lesson to actual students or to administer the PA task, the preservice teachers said they wanted to do both. They believed the performance task was valuable, and they learned from their experiences.

The planning guide served as a useful tool for preservice teachers in identifying their primary teaching and learning goals, and selecting and/or adapting a task around those goals. Inservice teachers could benefit from using such a guide for planning for teaching and learning with performance assessment. Additionally, inservice teachers might consider using the coding guide (Appendix C) adapted from Fuchs et al. (1999) as a framework for evaluating and adapting tasks for classroom use.

While preservice teachers grew in their understandings of performance assessments, there were details regarding partnerships between inservice and preservice teachers that needed more attention. Specifically, inservice teachers at appropriate grade levels and content areas and with necessary background knowl-

TABLE 1.

Selected Science Performance Assessment Tasks Designed for Intermediate and Middle School Students by the Preservice/Inservice Teacher Collaborative Teams

Task	Science Content
Design a Zoo: Consider habitats, animals, mapping, and costs.	Animal habitats
Ice Busters: Design a way to melt the ice on a coliseum floor so the 'N Sync band can play.	Melting, change of state
Blood Cells: Be a white blood cell or red blood cell. Write a "friend" about your travels through the body.	Body systems
Growing Plants: You are opening a garden center. Plan a way to keep your plants strong and healthy in your center.	What plants need to survive
Grocery Store Botany: Write an advertisement to get kids to eat vegetables, noting the role of eating leaves (blade and petiole).	Learn parts of plants
Weather and Evaporation: Write a report to water technicians to recommend the frequency of monitoring water evaporation under different weather conditions.	Cause and effect, change and constancy
Propagation: Seeds are not the only way. Write to nursery owners recommending methods for reducing their production costs by propagating plants in various ways.	Using seeds, cuttings, roots to create more plants; compare which methods give exact replicas of plants
NASA Food Weight Experiment: Explore ways of packaging food to get the most food aboard the shuttle.	Comparison of weight of multiple small packages versus fewer large packages
Density: Explore density of objects. Compare which float and sink. Write report to other scientists.	Compare densities of irregular objects
The Case of the Disappearing Water: Write a letter to the king telling him that no one stole his water—it was a part of the water cycle that it evaporated. Describe the water cycle in the letter.	Water cycle
A Visit to Jupiter, Saturn, Uranus, and Neptune: Students are astronauts visiting the planets and recording observations of their journey.	Identify and describe features of Jupiter, Saturn, Uranus, and Neptune
Scientist Discovers New Species of Insects: Scientists find new species of insects, build models that contain all body parts, and write reports about why they are insects.	Insects and their body parts

edge of performance assessment were needed. Better mechanisms for matching inservice teachers with preservice teachers also have to be developed.

The university instructors collaborated with ESD personnel and inservice teachers to develop a project that served not only preservice teachers, but also inservice teachers, to develop reform-based teaching practices. Through working with preservice teachers, the inservice teachers gained a better understanding of performance assessment—what it is, how it should be used, and how their own students participated in the tasks. Inservice teachers can look forward to improving their own teaching when they collaborate with preservice teachers on developing and administering PA tasks.

Another positive implication is that preservice and inservice teachers can both benefit from a collaborative project. The preservice teachers gained valuable information about working with children and designing and implementing PA tasks from the inservice teachers. The inservice teachers gained an appreciation and improved understanding of performance assessment from working with preservice teachers. This reciprocal relationship implies that a collaborative, goal-oriented project—situated in a classroom context that is meaningful for both parties—can provide benefits to both preservice and inservice teachers in understanding and administering science performance assessment tasks.

APPENDIX A
PEDAGOGICAL BELIEFS IN SCIENCE SURVEY/INTERVIEW QUESTIONS (Adapted from Peterson et al. 1989)

Describe, as specifically as you can, a lesson in which you introduce a new science topic to your class. We are interested in the way you organize and present the science content, as well as the specific teaching methods and strategies that you use. Preservice teachers: *imagine* a lesson and describe it (if you have not had experience teaching a new science topic). Inservice teachers: *recall* a particular lesson and describe it. How does your *introductory* lesson differ from a *typical* lesson on a science topic?

Describe, as specifically as you can, a lesson in which you include elements of the nature of science. We are interested in the way you organize and present the philosophy, as well as the specific teaching methods and strategies that you use. State specifically the elements you included. Preservice teachers: *imagine* a lesson and describe it (if you have not had experience teaching science). Inservice teachers: *recall* a particular lesson and describe it.

Describe, as specifically as you can, a lesson in which you include writing in science activities. We are interested in the role of writing in the lesson and the type of writing expected, as well as teaching methods and strategies that you use with writing. Preservice teachers: *imagine* a lesson and describe it (if you have not had experience teaching a new mathematics and/or science topic). Inservice teachers: *recall* a particular lesson and describe it.

What do you think the role of the learner should be in a lesson involving problem solving and reasoning?

Are there certain kinds of knowledge and/or skills in science that you believe all students should have? If so, what are they?

For the grade that you teach (or intend to teach), what do you believe should be the relative emphasis in science (1) on fact knowledge, (2) understanding scientific concepts and processes, and (3) solving of real-world problems? Why?

What do you see as the relationship between learning of scientific facts, understanding scientific concepts and processes, and solving real-world/authentic problems involving science?

What do you think the role of technology (e.g., calculators, computers, internet) should be in teaching and learning science?

Students have different abilities and knowledge about science. How do you find out about these differences?

Describe, as specifically as possible, what you understand performance assessment to be, when you believe it is useful, and when you believe it is not appropriate to use. If you have used performance assessment in your teaching, describe how you have used it.

Write and/or describe a science problem that might be categorized as an example of performance assessment.

Appendix B
Performance Assessment Task Mentorship Survey (to Be Completed by the Inservice Mentor Teacher)

Name_____School_____

Grade/Subject_____University Preservice Teacher_____

1. How would you define the term *performance assessment*?

2. How often do you do performance assessment in your classroom?
Often (weekly)___Sometimes (monthly)_____Seldom (1-2 times/year)____Never____

3. Have you had any classes/workshops/inservices on performance assessment? If yes, please describe.

4. Briefly describe your preservice teacher's performance assessment task

5. How would you rate the success of the preservice teacher's performance task implementation? (This will be held completely confidential.)

6. How well do you feel the students' task met the criteria for performance assessment?

   Did the task:
         a. present students with a task, project, or investigation? YES NO N/A

         b. establish a meaningful context based on issues/problems, themes, and/or students' ideas?
            YES NO N/A
         c. require the application of thinking skills/processes? YES NO N/A

         d. call for products/performances with a clear purpose for an identified audience? YES NO N/A

7. Did you learn anything about performance assessment tasks through this mentorship experience? If so, please describe.

8. Do you have any suggestions for the improvement of this mentorship project for the future? Please use the back of this form to complete this question.

APPENDIX C
CODING GUIDE FOR PERFORMANCE ASSESSMENT ELEMENTS PRESENT IN TASKS OR DESCRIPTIONS OF TASKS (Adapted from Fuchs et al. 1999)

Assignment: Write (describe) a math/science problem that might be categorized as an example of performance assessment.

Coding: Code "1" if present, "0" if not present.

_____ Contains two or more paragraphs
_____ Contains tables or graphs
_____ Has two or more questions
_____ Provides opportunities to apply three or more skills
_____ Requires students to discriminate relevant/irrelevant information
_____ Requires students to explain work
_____ Requires students to generate written communication

LINKS TO THE NATIONAL SCIENCE EDUCATION STANDARDS
The project described in this article relates particularly well to National Science Education Standards Assessment Standard C (NRC 1996)—that assessment tasks are authentic. In a performance assessment task, students take on roles and tasks within a context that makes the assessment meaningful. Additionally, performance assessments allow teachers to determine how well students can identify a worthwhile and researchable question, plan an investigation, execute the research plan, and draft a research report—in short, whether or not students can conduct scientific inquiry (pp. 98–99). Performance assessment can help teachers see how students apply knowledge, plan their curricula based on student knowledge, develop self-directed learners, and research their teaching practices (pp. 87–89).

REFERENCES
American Association for the Advancement of Science (AAAS). 1993. *Benchmarks for science literacy.* New York: Oxford University Press.

Darling-Hammond, L., and B. Falk. 1997. Using standards and assessment to support student learning. *Phi Delta Kappan* 79: 190–199.

Dickinson, V. L., J. Burns, E. Hagen, and K. M. Locker. 1997. Becoming better primary science teachers—A description of our journey. *Journal of Science Teacher Education* 8: 295–311.

Fuchs, L. D., K. Karns, C. Hanlett, and M. Katzaroff. 1999. Mathematics performance assessment in the classroom: Effects on teacher planning and student problem solving. *American Educational Research Journal* 36: 609–646.

Kelly, M. K., and J. B. Kahle. 1999. Performance assessment as a tool to enhance teacher understanding of student conceptions of science. Paper presented at the Annual Meeting of the National Association for Research in Science Teaching, Boston, MA.

McTighe, J. 1999, June. Performance task blueprint. In *Second performance assessment summer workshop manual*, ed. Washington State Office of Superintendent of Public Instruction. Seattle, WA: Office of Superintendent of Public Instruction.

National Research Council (NRC). 1996. *National science education standards.* Washington, DC: National Academy Press.

Peterson, P., E. Fennema, T. Carpenter, and M. Loef. 1989. Teachers' pedagogical content beliefs in mathematics. *Cognition and Instruction* 6 (1): 1–40.

Putnam, R., and H. Borko. 2000. What do new views of knowledge and thinking have to say about research on teacher learning? *Educational Researcher* 29: 4–15.

Shepard, L. 2000. The role of assessment in a learning culture. *Educational Researcher* 29 (7): 4–14.

Shymansky, J. A., J. L. Chidsey, L. Henriquez, S. Enger, L. D. Yore, E. W. Wolfe, and M. Jorgensen. 1997. Performance assessment in science as a tool to enhance the picture of student learning. *School Science and Mathematics* 97: 172–183.

Spector, B. S. 1999. *Bridging the gap between preservice and inservice science and mathematics teacher education.* Paper presented at the Annual Meeting of the National Association for Research in Science Teaching, Boston, MA.

Stenmark, J. 1991. *Mathematics assessment: Myths, models, good questions, and practical suggestions.* Reston, VA: National Council of Teachers of Mathematics.

Washington Commission on Student Learning. 1998. *Essential academic learning requirements.* Olympia, WA: Washington Commission on Student Learning.

Resources

Cavendish, S., M. Galton, L. Hargreaves, and W. Harlen. 1990. *Observing activities.* London: Paul Chaplin Publishing Ltd.

Doran, R., F. Chan, P. Tamir, and C. Lenhardt. 2002. *Science educator's guide to laboratory assessment.* Arlington, VA: NSTA Press.

Hubbard, K. M. 2000. *Performance-based learning and assessment in middle school science.* Larchmont, NY: Eye on Education.

Russell, T., and W. Harlen. 1990. *Practical tasks.* London: Paul Chaplin Publishing.

Schilling, M., L. Hargreaves, W. Harlen, and T. Russell. 1990. *Written tasks.* London: Paul Chaplin Publishing.

Stiggins, R. J. 1997. *Student-centered classroom assessment.* 2nd ed. Upper Saddle River, NJ: Simon and Shuster.

AUTHOR AFFILIATIONS

Valarie Akerson is an associate professor of elementary science education at Indiana University and has five years of elementary teaching experience. She has worked in preservice elementary science teacher preparation and inservice elementary science teacher professional development for six years.

Amy Roth McDuffie is an assistant professor of mathematics education at Washington State University and is a former secondary mathematics teacher. She focuses on preservice and inservice professional development toward standards-based mathematics instruction.

Judith A. Morrison is an assistant professor of elementary science education at Washington State Unitversity and is a former high school science teacher. Her area of focus is on informal diagnosis of students' understanding of science.

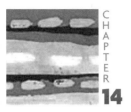

assessment in support of contextually authentic inquiry

Cory A. Buxton

EMPOWERING STUDENTS TO DOCUMENT THEIR OWN LEARNING

I found Ms. Johnson's class outside on the basketball court painting large canvases of handmade paper with dyes they were producing from native plant species as part of an integrated science and social studies unit entitled Mardi Gras Indians of New Orleans.

"Don't forget to use the sketches your group made last time," reminded Ms. Candice, one of the three science methods field experience teachers working with this group of 28 fourth graders, all of whom had been retained in grade this year for failure to meet the minimum criteria for passing the state high-stakes assessments. "Make sure you have a plan before you start painting!"

"And be sure you are filling in the data table up here for each color!" added Ms. Gladys, gesturing toward the large chart taped to the wall that bordered the court on one side. There were columns for characteristics of the dyes, including weight of botanical added, volume of water added, length of time cooked, final temperature, and sample of color.

Everyone seemed to be engaged in some aspect of the work. Some students were measuring and cooking the dyes under the supervision of the third field experience teacher, Ms. Tammy; others were using paintbrushes, sticks, and fingers to smear the brightly colored fruit and vegetable dyes onto their large sheets of homemade paper, attempting to mirror their pencil sketches.

Cedric, well known as one of the most challenging students in the class, came up to me, reaching for my video camera and insisting, "Lemme show you around the village."

I gave Cedric the camera and followed him as he took me from group to group filming each group's creation and asking them to describe what their canvas showed.

"Ours show how the Mardi Gras Indians make a new suit every year," reported Louis. "See, we got our beads and our feathers and the people be sewing and sewing. Here on this side is going to be a picture of the finished suit ready to march on Mardi Gras Day!"

While the paints dried, the class finished filling in the chart and discussing similarities and differences between the various dyes. Back in the classroom later in the day, students wrote up project reports, documenting the process they had gone through, discussing why each step in the process was necessary and reflecting on what they would do differently if they had another chance to repeat the activity. Ms. Johnson, the supervising teacher for the class, commented to me as the students wrote their reports that this was the most focused behavior she had seen from her class in weeks.

Later that week, when the preservice teachers returned to the elementary school, the students shared their murals with each other and with the preservice teachers and made a plan to also share them with a class of first graders the next day. At that point, Cedric inquired about the movie he had filmed. Ms. Johnson allowed Cedric and several other interested students to work with me editing the video they had taken to create a three-minute mini-documentary on the project, which they shared both with their class and with the class of first graders who had seen the finished products.

In her teaching journal, Ms. Tammy reflected on her experience with the project teaching, commenting, "I had some reservations about the mural activity, first that the nature of the project might lead to a lot of inappropriate behavior with this class of 'at risk' students and second, that they might have a lot of fun, but not learn anything. I think just the opposite was true. Their behavior was very good because they were enjoying what they were doing and they demonstrated to me that they learned a lot because they had a lot of ways to demonstrate what they had learned—orally, in writing, and even in video. There was a layered quality to the assessment that I found really appealing. We talked in methods class about contextually authentic assessment, but now I have a clear idea of what it looks like." (field notes)

RESEARCH CONTEXT

Tammy's comment about contextually authentic assessment highlighted the emphasis of this three-year research project on contextually authentic inquiry. This research project developed a model of science teaching, the contextually authentic inquiry (CAI) model, to guide both preservice and inservice teachers through inquiry-based experiences and assessments as active participants in problem posing and problem solving. Teachers were being prepared to use this model of inquiry and assessment with their own students.

The overarching question focused on what it means to make learning—and the assessment of that learning—"authentic" for students and for teachers in academically struggling, urban elementary schools. This issue (of authentic teaching, learning, and assessment) also must be considered within the broader policy context of standards-based assessment and accountability.

RESEARCH QUESTIONS

The following three research questions (part of a larger study) were investigated:

1. What resources are required to support the engagement of teachers and students in contextually authentic science learning experiences?
2. Can contextually authentic science learning experiences be used to support teachers' desires to improve students' canonical understandings of science (i.e., the traditionally accepted Western scientific norms)?
3. What does it mean for teacher educators and teachers to share responsibility for improving school performance in a low-performing urban school setting?

COLLECTION AND ANALYSIS

Data collection focused on the actions and interactions of groups of participants in elementary schools. The data were composed of the following:

* Ethnographic field notes taken by the researcher and a graduate assistant for each class session attended. These field notes were both a record of what transpired and the researcher's on-the-spot reflections and interpretations of those events.
* Assignments and products from the students and teachers involved in the teacher education program.
* Photo and video recordings of selected class activities by both the researcher and students.
* Practicum journals and surveys from preservice teachers in which they described positive and negative aspects of their practicum teaching experiences and reflected directly on research questions.

Data were first coded into cultural domains using both the sociolinguistic categories proposed by Spradley (1980) and the conceptual categories that emerged as the researcher constructed and refined the inquiry model. Second, these domains were arranged into a taxonomy that informed future data collection, additional domain coding, and model revisions. Third, the domains were organized into tables along dimensions of contrast, where each data set was revisited to identify examples of contrasts across domains. Fourth, the tables were used to generate assertions that could be related to the research questions and then supported with specific examples.

Semantic structure analysis provided a systematic and rigorous approach to analyzing the various data streams (illuminating cultural behaviors, cultural knowledge, and cultural artifacts) in order to examine how participants viewed and made sense of their own interactions. In contrast, vignette analysis (Graue and Walsh 1998) provided an opportunity to explore, and express in a holistic way, the researcher's developing understanding of the cultural scenes studied.

FINDINGS

In the case of Jemison Elementary (the setting for the vignette at the beginning of this chapter), there was a palpable policy tension between teachers' interest in exploring various meanings of authentic teaching and assessing and teachers' concerns that the high-stakes testing context dictated a highly scripted approach to classroom instruction and assessment (e.g., all assessment should be modeled after the state test). Over the two and one-half years of this study, the teachers and students at Jemison Elementary, located in a Southeastern urban center, felt an ever-increasing pressure from the state assessments. As a researcher, I observed frequent discussions among teachers that reflected a feeling of disempowerment, often bordering on marginalization and hopelessness. Although students were less articulate about their anxiety, when the topic of the high-stakes test was raised, their body language reinforced the perception of tension. To combat this sense of marginalization, many teachers wished to promote approaches to classroom assessment that they perceived as being more empowering to students. A common thread was their desire to give students more control over documenting their learning. I began to routinely ask teachers the question "What can you do to help students take increased ownership over documenting what they have learned?"

FRAMEWORK FOR INTERPRETATION

Fostering authentic learning and assessment experiences naturally raises questions about what qualifies as an authentic learning or assessment task. Does authentic assessment have to do with specific types of assessment activities, such as portfolios or lab practicals? Does it have to do with connections between how the material in question is taught and how it is assessed? Does authentic mean that a teacher is attempting to make a connection between what his or her students find personally meaningful about what is being taught and how that learning is to be assessed?

There are three possible models for authentic science inquiry (Buxton and Whatley 2002). The first model, "canonical authenticity," highlights a view of science concepts and processes modeled after "scientists' science," the science that takes place in the research lab (Chin and Malhotra 2002). From this perspective, assessment invokes the reasoning that scientists employ as they engage in inquiry. The second model, "youth-centered authenticity," takes as authentic those tasks that start with the interests, perspectives, desires, and needs of the student (Barton 2001). From this perspective, assessment permits youth to explore and make use of science and technology for their own purposes. The third model, "contextual authenticity," combines the knowledge, skills, and habits of mind promoted in the canonical model with the linguistic, cultural, and identity features that are valued in the youth-centered model. In the contextual authenticity model, assessment must foster the investigation and clarification of the intersections between students' everyday knowledge and canonical scientific practices in ways that empower the student to participate in inquiry.

EMERGING PRINCIPLES

Using this framework with the teachers and students at Jemison, four principles emerged as common characteristics of contextually authentic assessment. These principles, and what they looked like in practice across the three groups of participants (inservice teachers, preservice teachers, and elementary students), are outlined in Table 1.

TABLE 1.
Four Principles of the Contextually Authentic Inquiry Model and How Teachers and Students Responded to and Enacted the Principles

Guiding Principle	How Inservice Teachers Responded to and Enacted the Principles	How Preservice Teachers Responded to and Enacted the Principles	How Elementary Students Responded to and Enacted the Principles
1. Connections to students' prior knowledge and experiences	Interested in assessing what students knew rather than what they did not know.	Wanted to assess students using a variety of methods, not just forms that model the high-stakes test.	Wanted greater control over documenting their learning.

Had strong desire to use technologies and multimedia.

Valued taking all assessments seriously as way to prepare for high-stakes tests (emerged during the project). |
| 2. Increased student choice within a wide range of assessment methods | Resisted (somewhat) student use of technology for instruction and assessment. | Learned to differentiate between assessment and grading.

Valued beginning instruction with a deep understanding of student prior knowledge (emerged during the project). | |
| 3. Active use of educational technology | Valued assessment as formative part of the learning process (emerged during the project). | | |
| 4. Assessment as an organic part of the learning process | | | |

The first principle of contextually authentic assessment, connections to students' prior knowledge and experiences, was clearly expressed by Ricki, one of the preservice teachers who completed her field experience:

These teaching experiences made it really clear for me how all students learn differently and therefore assessment does not have to be the same for each student. In the past I believed that assessment had to be in the form of a written test and that it was only fair if it was the same for everybody. After seeing firsthand the various ways that students learn and express their knowledge, I believe that there can be different types of assessment for different students at different times. Just as instruction needs to take into account students' prior knowledge and experiences, our assessment needs to do that too. By using a variety of assessment methods, I feel that I will be able to better gauge the students' learning and comprehension of what I taught. And by looking for teaching and assessment strategies that connect the standards to my students' experiences, I feel that I will be able to get my students to perform at their best. Hopefully, that will be enough for them to pass the [high stakes tests].

(teaching journal, Ricki, 4/9/02)

Ricki's reflection related to the second principle of contextually authentic assessment, *increased student choice within a wide range of assessment methods*. Many low-performing students expressed confidence and competence in a range of skills, including singing, sports, dance, use of media (such as music and video), even as they expressed discomfort and a lack of confidence in their abilities to perform traditional academic tasks (such as standardized tests). As Shantel, a student in Ms. Johnson's class, eloquently described her feeling, "That test is all about showing what I don't know. When is somebody gonna ask me about what I do know? They shouldn't be giving us tests that make us feel stupid!" (field notes, 4/18/02). The group of preservice teachers working in Ms. Johnson's class came to see multimedia approaches as examples of contextually authentic assessments that played into student strengths, rather than their weaknesses. Other contextually authentic assessment activities that students used to express what they learned were music, poetry, rap, and visual arts.

This connection to multimedia highlighted the third principle of contextually authentic assessment, the *active use of educational technology*. Digital moviemaking provided a powerful assessment tool for both preservice teachers and students at Jemison. These movies served as learning experiences in their own right, as a tool for documenting what an individual or group learned, and even as a cultural critique—a way to present images that combat certain narratives and stereotypes of urban schools and students. Movies and visual images can be powerful because they change the way we relate to the world. They can serve as powerful assessment tools in this same way. Hayes (2001) argued that students come to see moviemaking projects as a form of research that requires collaboration and negotiation, in contrast to the isolated and isolating task of writing a paper or taking a test. The difference in the classroom is a question of audience. Students view a paper or a test as having an audience of one—the grader—but a movie potentially has a broader audience. Students often seem driven to make their movies as high quality as possible.

The use of moviemaking as both a learning experience and a vehicle for documenting learning led to the final principle of contextually authentic assessment, *assessment as an organic part of the learning process*. As growing significance has been placed on standardized test scores, students (especially those who are perceived as being at high risk of failing these high-stakes tests) are likely to be subjected to classroom instruction and assessment meant to simulate these tests. In such cases, assessment remains a construct separate and apart from the learning process. A focus on what makes learning contextually authentic for students provides a fundamentally different way to relate classroom instruction and assessment. While not a new pedagogical idea, this focus creates a clear bond between instruction and assessment. Making assessment part of the learning process is an approach that seems to have been largely lost in the push to prepare urban students to pass high-stakes tests. As Ms. Johnson commented, "If we can get the kids to see assessment as something they naturally do all the time to show us what they've learned, I'm convinced that when it comes time for the test they won't be as stressed."

CONCLUSION

As a teacher educator, I was committed to exploring features that helped or hindered teachers' attempts to implement the contextually authentic inquiry model with their students. Working with the model taught me about assessment with children in urban schools, where students are expected to be at risk of academic failure. If research into the assessment of science learning will make a difference for teachers, it must be applicable to the realities of K–12 classrooms. Those of us who conduct our research in urban schools must especially take care that the reforms we are promoting are feasible for teachers and students, given the resources and multiple challenges of teaching in urban settings. At the same time, we must be careful not to sell students (and teachers) short in terms of our beliefs about their capabilities.

This research project confronted me with a dilemma as a teacher educator, namely, how to implement a model of authentic inquiry-based science within a culture of high-stakes assessment. My research at Jemison led me to believe that such an approach is possible, but only if teachers are convinced of its value. If the current "bottom line" for teacher accountability is a measure of student success on high-stakes tests, then models of contextually authentic teaching and assessment have to document that they are not detrimental to student performances. While I believe this to be the case, more research is clearly needed to focus on these kinds of student outcome measures. My experience working with teachers and administrators in low-performing urban schools indicates that school policy tends to fall back on teaching to the test. This situation represents a static and conceptually disconnected model of assessment that is not authentic in any of the senses outlined previously. As students in urban schools continue to fail the high-stakes standardized tests in greater numbers, it is clear that our current assessment practices are not working.

As an educator who seeks teaching and assessment strategies that will help urban students in under-resourced schools gain the skills and knowledge needed to pass these high-stakes exams, I have come to the conclusion that contextually authentic models of teaching and assessment offer promise. Approaches that aim to connect what is meaningful to students (and teachers) with what is mandated that students (and teachers) learn about science can only benefit everyone as we create classroom communities to promote scientific literacy for all.

LINKS TO THE NATIONAL SCIENCE EDUCATION STANDARDS

One of the biggest challenges science teacher educators face is to raise the awareness and thinking of preservice and inservice teachers about standards and benchmarks in a systemic way. Because of states' emphasis on content standards over other standards, teachers tend to overlook the more holistic intent of the National Science Education Standards (NRC 1996). While my teaching and research address national and state *content* standards, I place an equal emphasis on the other National Science Education Standards. For example, in my classes preservice teachers address teaching standards such as planning an inquiry-based program and developing a community of science learners. We consider the importance of professional development standards (such as gradually improving one's knowledge of essential science content and learning to integrate content knowledge with pedagogical knowledge). My classes always focus on the science assessment standards, especially matching the assessment type with the learning task.

REFERENCES

Barton, A. C. 2001. Science education in urban settings: Seeking new ways of praxis through critical ethnography. *Journal of Research in Science Teaching* 38 (8): 899– 917.

Buxton, C., and A. Whatley. 2002. Authentic environmental inquiry model: An approach to integrating science and social studies in under-resourced urban elementary schools in Southeastern Louisiana. Paper presented at the Annual Meeting of the American Educational Research Association, New Orleans, LA.

Chinn, C., and B. Malhotra. 2002. Epistemologically authentic inquiry for schools: A theoretical framework for evaluating inquiry tasks. *Science Education* 86: 175–218.

Graue, M. E., and D. J. Walsh. 1998. *Studying children in context: Theories, methods and ethics.* Thousand Oaks, NJ: Sage.

Hayes, M. T. 2001. Science is over at 1:35: Performing the elementary science curriculum as disciplinary space/time. Paper presented at the 2001 meeting of the American Educational Research Association, Seattle, WA.

Spradley, J. 1980. *Participant observation.* Fort Worth, TX: Harcourt Brace Jovanovich.

RESOURCES

Apple Learning Exchange: Teaching and Learning
  Explores the potential of the digital classroom including technologically innovative approaches to teaching, learning, and assessment. An excellent source of ideas for lessons that make use of technology. The website is *http://ali.apple.com/ali_sites/ali/teaching.html*.

Urban Science Education Center, Teachers College, Columbia University
  Promotes equitable science learning experiences for students in urban schools, with resources and research on a range of topics related to urban science education. The website is *www.tc.edu/centers/urbanscience*.

AUTHOR AFFILIATIONS

Cory A. Buxton is assistant professor of science education in the School of Education at the University of Miami. He received his PhD from the University of Colorado at Boulder in 2000. His research interests involve the interactions of culture, language, and tool use in the teaching and learning of science in urban classrooms. He works on a National Science Foundation-funded project entitled Promoting Science Education for English Language Learners (P-SELL), examining how teachers' reflections on students' reasoning influence the ways in which teachers approach science instruction and assessment with English language learners.

helping students understand the minds-on side of learning science

Lawrence B. Flick and Michael Tomlinson

INTRODUCTION

Elementary students are usually very eager to become involved in activities. When asked about what makes something interesting and what more they would like to know, however, the discussion may fall flat. Students will repeat an interesting result a dozen times, but balk at reflecting on the experience and describing or discussing its scientific meaning. This instructional problem brought Flick, a former middle school teacher turned science educator, together with Tomlinson, an elementary teacher and Christa McAuliffe Fellow. They shared an interest in connecting research on teaching reading with research on learning science.

Poor readers can be supported in the use of cognitive strategies used by skilled readers (Brown and Campione 1998). These cognitive strategies can be tailored to other tasks such as inquiring into scientific problems (Brown and Campione 1990). From this point of view, students are not so much refusing to reflect on science activities as much as they may be lacking the skills for doing so (Bransford, Brown, and Cocking 2000). This idea launched an action research project by Tomlinson to evaluate how the teaching of cognitive strategies improved student performances on science inquiry assessment tasks. This chapter focuses on the first two steps in this project: (1) the selection of the target strategies and (2) designing methods for teaching target strategies to fourth-grade students.

The National Science Education Standards have challenged the nation's science teachers to teach science as inquiry (NRC 1996). The state of Oregon has stipulated that, as part of the state assessment system, teachers design inquiry tasks and score student work samples using a state scoring guide. To meet this challenge, Flick and Tomlinson began a process of examining the literature on cognitive strategy instruction in reading comprehension.

Dole et al. (1991) reviewed two decades of research on reading instruction, describing the transformation from teaching discrete reading skills (e.g., topic sentences, main idea, and sounding out words) to teaching cognitive strategies used by skilled readers. Skills are thought of as nearly automatic behaviors used in a routine fashion. Strategies, by contrast, are conscious and flexible plans called into action by readers and adapted for a variety of contexts—for example, "employ the broad strategy of 'determining importance' to accomplish comprehension tasks" is a typical strategy. There is far more involved with this strategy than automatically using the skill of finding a topic sentence. Good readers often adjust their purpose for reading to sort information from different points of view and to accomplish different tasks. Often confused with determining importance is the strategy of summarizing information. Summarizing is a more powerful strategy than the skill of identifying important information. Summarizing requires the reader to use the salient information in a synthesis that succinctly captures big ideas, perhaps highlighting an order or sequence that makes the meaning of the text clearer.

Tomlinson applied these strategies across the curriculum. The following classroom scenario illustrates a cognitive strategy instruction with fourth-grade students who are learning about scientific inquiry. (Tomlinson's students refer to him as Mr. T.)

BACKGROUND: A CLASSROOM SCENARIO

Mr. T's fourth-grade students have been engaged in activities with circuits. They used a battery, bulb, and wire to light the bulb. Students were prompted to manipulate the bulb and light it in different orientations. When challenged to use other materials to light the bulb, one student found a way to hold the battery on a notebook ring and used the wire to light the bulb. This idea spread quickly, and soon students were using a variety of materials to complete the circuit. This activity had never failed to spark the interest of students. Mr. T was familiar with how students discarded nonconducting material and focused on metal objects. The more adventurous students used materials they were wearing, such as belt buckles, glasses, or braces.

This class was similar to one Mr. T had run several times before. This time, however, there was a difference. In previous efforts, Mr. T had been frustrated when he prompted students to reflect on their experiences with circuits and raise questions and develop their own investigations. The electric circuit activity was always ripe with possibilities for fourth-grade, student-designed and -conducted investigations. However, in the past, only a few students would engage with the hands-on experiences beyond the initial instructions. Getting students to think beyond these initial activities, to questions that would lead to an investigation, often had resulted in a long string of teacher prompts. Mr. T had ended lessons feeling as if he were the one doing all the thinking.

Mr. T instituted a program of instruction to develop the use of key cognitive strate-gies designed to support higher-level thinking. These strategies were posted on the wall and the terms themselves were integrated into discussions with various organizers used across the curriculum. The cognitive strategies were selected after consulting re-search findings on reading comprehension, inquiry science, math problem solving, and literacy instruction. The strategies were (a) setting a purpose, (b) using prior knowl-edge, (c) looking for patterns, and (d) metacognition.

Each day began by setting goals and discussing the purpose of each activity. Al-though students sometimes had trouble saying "metacognition" and were not sure of the exact meaning of the word strategy, they were clear on what was expected of them by each term. Mr. T asked the class, "What does it mean to set a purpose for your investigation?" A typical response was, "You think about what you are doing. It would help you decide what you need to be doing." Another student added, "Like it helps you recognize when you are off task. If you are off task, you need to recognize it and say what you need to do to get on task again."

One day, the purpose was to develop a question and a plan for a science investiga-tion. Students engaged in several hands-on activities to gain familiarity with the topic of electricity. Mr. T prompted the students to identify variables and use these variables to construct questions. After Mr. T led a warm-up, small groups of students brain-stormed lists of questions. Mr. T prompted students to think of prior knowledge about and experiences with electricity (e.g., earlier activities with battery and bulbs, per-sonal experience with electricity around the home or in the community, and knowledge gained from TV or other media). Individually, students evaluated their brainstormed lists for questions to guide their investigations. Mr. T prompted students to recall the purpose: to identify a plan for scientific investigation at key points during the activ-ity. Mr. T understood that providing supports or scaffolding for student higher-level thinking meant that he had to model, create structures, and offer reminders about was meant by, in this case, "setting a purpose."

Each student had a graphic organizer that had been designed by Mr. T to support his or her thinking during the development of his or her investigations. The graphic organizer guided students in summarizing exploratory hands-on activities and listing materials, procedures, observations, and variables. Each student wrote the purpose at the top; one student wrote, "I want to investigate something interesting about electric circuits." Her group brainstormed a list of five questions: Why does the power go off sometimes? How many lights can you light at once? How come some lights are brighter? Where does electricity come from? How do you make an electric circuit?

As students worked individually to select a question, they used prompting questions from an organizer for the day's lesson: (a) Why is the selected question important?, (b) Why did you select this question?, and (c) How does this question relate/connect to your purpose? To support this work, students also consulted other print and material resources in the room. Some students read from books, others gathered around the one computer available for classroom use and checked the internet, and others examined electrical devices including loose batteries and bulbs.

Mr. T told students to pause in what they were doing and record their thoughts in a Thinking Log. The Thinking Log was a set of blank pages, folded and stapled in the middle, with an outside cover personally decorated and labeled by each student. Students understood the Thinking Log was a metacognitive prompt. Mr. T recognized that students were building metacognitive skills and needed structured guidance to reflect on their work. In setting up the Thinking Logs, Mr. T had students write a series of sentence stems as thinking starters inside the front cover. These stems served as prompts for log entries, whose purpose was to prompt metacognitive awareness of confusion or lack of understanding. The set of stems included the following examples:

I got confused when...so I...
I got stuck when...
I stopped...because...
I figured out that...
I first thought...then realized...
This reminds me of...
I'm not sure...
I didn't expect...
I think I need to rethink, redo, reread...
To better understand...I need to learn more about...

Mr. T circulated in class as students wrote. He gave feedback (formative assessment) to students about the content of their entries (e.g., "Try to be as specific as you can about what you are confused about."). He also made a mental note of entries that would be valuable for sharing with the entire class. Mr. T asked for feedback from the students after they finished writing. He used the examples he had mentally noted from student papers in case students were reluctant to share their thinking. This feedback time gave students another opportunity to build metacognitive skills by hearing how others expressed their thinking.

With the purpose statement at the top of the graphic organizer and the question for investigation recorded below, students continued consulting classroom resources. They made notes in the organizers with one column headed "Information From Text" and a second column headed "Connections: Prior Knowledge." Mr. T prompted students to think about key ideas from reading they had been doing in classroom resource books. For example, one student wrote in the Information column: "Power poles carry wires from the power plant to homes. Many homes can be powered from the same wires." In the Connections column he wrote, "I used wires in my circuit. Lamps have wires." After students listed 10 statements from their reading, they came to a space at the bottom of the page labeled "Inferences." Mr. T guided students in making inferences by asking students to "relate text to text, text to world, and text to self." This instructional strategy helped students check their own comprehension. Relating text to text *meant students examined how the different materials they read related to one another. Relating text to world meant "How does what I am reading relate to my world and the world in general—in other words, to my prior knowledge?" Relating text to self meant "How does what I am reading relate to what I know and my experiences?" This strategy prompted metacognitive thinking and helped students see how their reading and personal knowledge related to their science investigations. In one case, a student wrote, "Wires are used for electricity." This student would have to evaluate the information gathered relative to his question for investigation: "What makes bulbs light brighter?"*

The lesson ended when Mr. T asked students to share the questions they had selected for their investigations. Some students had already begun to design procedures for their investigations, using the materials available in the room. Selected students shared their investigation questions and ideas for procedures with the whole class. Mr. T's closure was to begin discussion of what should happen tomorrow, as a way of setting a purpose for tomorrow's activity. Students suggested that they should finish designing procedures and decide what materials they would need. Mr. T prompted students to explain their ideas to a partner. He made a quick check to see which students were having trouble verbalizing ideas and made a note to meet with them during the break. The lesson closed with students placing their graphic organizers into their science folders.

APPLYING RESEARCH TO INSTRUCTION: COLLABORATION BETWEEN COLLEAGUES

The previous classroom scenario highlighted how Tomlinson applied research on teaching cognitive strategies. Dole et al. (1991) distinguished between teaching individual skills and teaching cognitive strategies. Traditional reading and science instruction has often focused on skills (e.g., finding the main idea, using context to find word meaning, and, in science, graphing data and making observations or inferences). Tomlinson had previously focused on the mechanics of making and drawing circuits. Now his emphasis was on setting and maintaining a clear purpose for what could be learned about electricity. Rather than simply getting a bulb to light in various ways, students were prompted to look for patterns in the combination of materials that lighted the bulb(s). He prompted students to connect what they knew about electricity to the current experience, as opposed to narrowly focusing on discrete events of the science activity.

Using the concept of cognitive strategy as a guide, Tomlinson and Flick shared ideas and research articles to select a set of teaching strategies to support learning in general and high-level tasks in science in particular. From the standpoint of a classroom teacher, this approach offered Tomlinson a way to think about how to help students understand and stay focused on the major task, without their being sidetracked by procedural details. From experience, Tomlinson knew that the most difficult aspect of guiding student inquiry about a science problem was to generate and maintain focus on the problem. Maintaining a sense of purpose was also a recommendation from research findings on reading comprehension.

The broad goals of reading comprehension and conducting scientific investigations seemed to be complementary within the context of the fourth-grade curriculum. In the early stages of their collaboration, Tomlinson and Flick wrote an article for the state science teacher journal (Flick and Tomlinson 2001), in which they created a figure describing what they saw as teachable elements in both reading and science inquiry (see Figure 1). They did not view the strategies as being in a one-to-one alignment, but as complementary parallel sets. This suggested to Tomlinson a way to think about high-level thinking goals in science as a part of his instruction—if students were developing higher-level cognitive strategies in science, then instruction in the other curriculum areas must complement and support the higher-level thinking and reinforce key strategies, such as determining or setting a purpose.

The four cognitive strategies and instructional approaches Tomlinson used are described below (Cook and Mayer 1988; Dole et al. 1991; McNeil 1987; Wittrock 1998).

Strategy #1: Setting a Purpose. Successful learners in general (good readers) tend to set goals. Extrapolating from expert versus novice studies, successful learners need to have a clear, accurate purpose before starting a task. The task's purpose helps the learner determine which skills and strategies to apply, while providing a meaning or context for the task. A strong sense of purpose focuses attention on the key ideas and details, increasing reading comprehension or developing student understanding of the scientific problem.

Tomlinson used explicit prompts to guide student thinking, beginning each day by guiding students in setting learning goals that were based on daily activities. Graphic organizers enabled students to compose a purpose statement. Tomlinson posed reflective questions such as, "Why did you select this question as most important?" or "Why was this prediction selected?"

Strategy #2: Using Prior Knowledge. Prior knowledge is the student's aggregate background of content knowledge and personal experiences that are brought to any task. When prior knowledge is activated before a task, it has the effect of validating the student's experience and knowledge while empowering the student and increasing self-confidence. Prior knowledge aids students in decoding text, which increases their fluency in expressing ideas. The degree to which prior knowledge is actively employed by a student will affect his or her ability to elaborate on, judge, and evaluate the reading text or scientific problem. Tomlinson modeled the use of prior knowledge by thinking aloud in class and then asking students for their responses. Effective strategies for activating prior knowledge included drawing "webs" of knowledge, K-W-L, concept maps about a specific idea, Venn diagrams, and making prereading predictions about what students expect the text to say.

FIGURE 1.

Teachable Elements of Scientific Inquiry and Reading Comprehension Applied to Science

Teachable Elements of Scientific Inquiry	Teachable Elements of Reading Comprehension Applied to Science
Problem Identification. Skilled science students use background knowledge in science and direct observations of the world, often stimulated by a discrepant event, to identify a problem for investigation. They see the problem in terms of identifiable factors and their relationships. Relationships may be descriptive, correlational, or cause and effect.	*Determining Importance.* Skilled readers have a purpose for reading that helps determine what is of most importance. They do this by using general knowledge about the world and science-specific knowledge. They consider the intentions of the text writer for conveying meaning. They use knowledge of text structure to recognize how information is organized.
Finding and Using Information; Facts; Observations. Skilled science students seek background information from textbooks, libraries, computer databases and networks, and commercial media and by talking to experts and to each other. Students also gain information from direct observations and distinguish these observations from inferences they make from the observations.	*Monitoring Comprehension.* Skilled readers spend mental energy on monitoring, controlling, and adapting their thinking as they read. Problems of comprehension, such as finding information inconsistent with their own preconceptions, are caught as they read. They employ additional resources where needed, and plan how to use resources to correct problems and achieve their purpose.
Applying Procedures and Skills. Skilled science students recognize that new procedures require learning new skills. They see procedures as a means to gaining and applying new information. To investigate a problem, students select from their repertoire of skills and design procedures.	*Summarizing.* Skilled readers compose summaries for themselves to aid in comprehension as well as in the preparation of summaries for school assignments. They produce summaries by intentionally selecting important information and ignoring unimportant information. Some information must be condensed and some is organized by using more inclusive concepts. The summary is then composed as a coherent and accurate representation of the original material.
Drawing Inferences. To skilled science students, *data* always has some meaning beyond merely collecting observations and completing procedures. From the data, students distinguish observations and inferences about particular events set up by the procedures. Students draw inferences from the observations to generate descriptions, trends, or relationships.	*Drawing Inferences.* Skilled readers construct their own model of text meaning by filling in details and elaborating where more information will aid their understanding. Readers use background knowledge and seek information from other sources to fill in perceived gaps in what they are reading.
Generating Interpretations, Explanations, and Applications. Students interpret data and create an incomplete and tentative explanation of events related to the original problem. Students consider how the explanation may apply to other problems. They can design another investigation based on their results.	*Generating Questions.* While teachers routinely ask questions about assigned readings, skilled readers generate their own questions to improve comprehension. These questions may be about the science content itself, the way the material is presented, or about how the information is useful.

Strategy #3: Looking for Patterns. Looking for patterns in nature is central to any scientific investigation. Pattern recognition strategies support comprehension of nonfiction, structured, and/or expository texts (including key ideas, details, relationships among ideas in the text, and relating new ideas to prior knowledge). Seeing an organizational pattern helps the learner make sense of printed material. Stories are patterns that help readers recall information by placing the information in a meaningful context. The same is true in science. Perceiving a pattern in the kinds of materials that will complete a circuit helps students not only remember characteristics of conductors (e.g., metal), but also remember characteristics of nonconductors.

Strategy #4: Metacognition. Metacognition refers to two distinct areas of research: (1) knowledge about cognition and (2) regulation of cognition (Bransford, Brown, and Cocking 2000). Tomlinson's instructional approach helped students understand what the four cognitive strategies were and why developing them was valuable for learning. One method he used was to give each strategy a specific descriptor. Students learned that *metacognition* was "thinking about your thinking." Through graphic organizers, questioning routines, and Thinking Logs, Tomlinson scaffolded learning situations that built student understanding through cognitive strategies and helped students to use the strategies in meaningful learning contexts.

ASSESSMENT
These four techniques were used for assessment:
1. The Cognitive Strategies Inventory (following Blumenfeld 1992) (student interpretation of classroom tasks)
2. A Thinking Log and a Reading Response Journal (students recorded reflections and ideas)
3. Graphical organizers (structuring devices for student engagement in high-level learning activities, such as math problem solving and science inquiry)
4. Interviews with the entire class (to determine student understanding of the cognitive strategies and how they were emphasized through instruction)

The Cognitive Strategies Inventory (CSI) was designed by Flick and Tomlinson and validated by a third science educator against the target cognitive strategies. The CSI contained four cognitive strategies and a separate list of skills (writing, discussing, computing, reading, public speaking, and problem solving). Students checked all skills and strategies that applied to the task in which they were engaged. The section "Evaluating my own work" asked students to mark a list of questions that applied to the current task (e.g., Did I do what was expected of me? Have I done my best work?). Students completed the inventory in less then five minutes and resumed their work.

Thinking Logs and Reading Response Journals were booklets of standard size paper folded and stapled in the fold. Students individually decorated and identified their booklets, and kept them in their desks. Unannounced, Tomlinson asked them to record what they were thinking and then evaluate their work. For example, students wrote in their Thinking Logs during writing assignments and math problem solving or when engaged in a science activity.

Students were interviewed in pairs about cognitive strategies. Responses from a pair of students tend to be more comprehensive than they would be from a single student; students prompt each other to remember various elements of instruction. Tomlinson used his judgment to pair students by their abilities to articulate their thinking, keeping one student from dominating the responses and maintaining the comfort level during the interview.

RESULTS

STUDENT INTERVIEWS

During spring 2002, Flick conducted interviews of students in Mr. T's class. All student pairs remembered the four cognitive strategies by name with little or no prompting. For example, a student would say "Setting a purpose keeps you on task; it lets you know what you are supposed to do." "Using prior knowledge" was defined in a circular fashion—"it means what you know" and was sometimes referred to as "making connections." This latter definition came from Tomlinson's instructional strategy of asking students to connect text to themselves, text to text, and text to world. When asked for an example of a pattern, students most commonly responded with a number series, such as 2, 4, 6, 8.

Although students remembered the strategies, some students had problems giving clear examples of what they meant. This contrasted somewhat with observations from the classroom. Students who responded with an explanation in class were usually the most capable students. Metacognition was generally described as "thinking about thinking." When asked, "What do you do when you are thinking?" most student pairs drew a blank. Students had no language for talking about their own thinking. When asked, "What did Mr. T do?", they referred to Thinking Logs.

There was an overlap in what students identified as examples of cognitive strategies. A given instructional method prompted more than one cognitive strategy; however, students were not able to go beyond explaining metacognition as staying on task and completing the assignment. They used a circular chain of reasoning: "Setting a purpose keeps you aware of what you are supposed to do. Metacognition reminds you that you were thinking of recess rather than the assignment. Being reminded of the purpose helps you stay on task so that you can get done and go to recess." This reasoning was similar to that found in other research studies. In a study of 275 fifth- and sixth-grade students with five different teachers, Blumenfeld (1992) found that students over-emphasized task completion and under-emphasized task understanding. The fourth graders in Tomlinson's class stated the same narrow purpose in new terminology, as seen in the following interview excerpts (interview questions are omitted; "S" indicates a student making a response to an interview question).

S: *He has this little Thinking Log…we have to write down what we are thinking.*

S: *You write what you were thinking at that time. And if it was something that was off task, like snack, and we were working on science, you'd probably suggest a fix to get your learning on track.*

S: *[A fix would be] to finish up your work quick so that you could start thinking about, like snack.*

S: *[Metacognition helps us learn] by helping us think about our thinking so that we can stay on task, keep our thoughts about what's going on and stuff.*

The five classrooms in the Blumenfeld (1992) study were observed 30 times each. From these data, she inferred that the disconnect between cognitive demand of a task and cognitive output by students resulted from (a) task length, when students could become fatigued or sidetracked and (b) the high-level thought required by self-discipline in reflecting, use of resources, or cooperation. These findings were also characteristic of the tasks observed in Tomlinson's classroom. While students tended to emphasize task completion, Tomlinson's instructional approach made cognitive strategies explicit. Students remembered and discussed off-task behavior and employed "fix-it" strategies (Dole et al. 1991). Skilled learners were not only aware that they were having a problem, but knew how to resolve it. Tomlinson prompted students to check on their learning progress and to consult a list of strategies for solving problems or ways to get new ideas (such as discussing a problem with a neighbor).

COGNITIVE STRATEGIES INVENTORY
We designed the Cognitive Strategies Inventory to monitor student understanding and use of the target strategies. Across the school year, Tomlinson administered the inventory during activities in which students were expected to be engaged in higher-level thinking. Students marked skills they had used during the preceding activity, then marked the cognitive strategies employed (see Table 1). Reading, writing, and discussing were the most commonly cited skills used for assignments. Students may not have differentiated public speaking from discussion. Sharing ideas in science class appeared to be identified by students as public speaking. (There was no formal public speaking component.) Problem solving was cited less often, and was generally associated with math problems at the beginning of the class. After Tomlinson explicitly highlighted problem solving in reading and science, students interpreted the term more broadly. Relating prior knowledge, setting a purpose, and metacognition ("thinking about my thinking") were the most commonly cited strategies. Looking for patterns and summarizing were least cited (students had difficulty stating what a pattern meant during interviews).

On the section of the inventory labeled "evaluating my own work," students

TABLE 1.
Cognitive Strategies and Academic Skills Marked by Fourth Graders

	Areas in Which Student Responses Were Collected					
Responses Collected on	Reading for Information **Oct. 31**	Math Problem Solving **Jan. 14**	Science Project **Jan. 24**	Science Project **March 6**	English Essay **March 13**	Range
Skills I used for this assignment:						
Writing	68% (17)	90% (19)	80% (20)	100% (20)	76% (16)	68–100%
Discussion	64% (16)	10% (2)	28% (7)	75% (15)	38% (8)	10–75%
Computing	0% (0)	29% (6)	0% (0)	40% (8)	5% (1)	0–40%
Reading	88% (22)	86% (18)	72% (18)	75% (15)	52% (11)	52–86%
Public Speaking	0% (0)	0% (0)	0% (0)	75% (15)	0% (0)	0–75%
Problem Solving	24% (6)	81% (17)	8% (2)	35% (7)	14% (3)	14–81%
Strategies I used for this assignment:						
Relating prior knowledge	68% (17)	67% (14)	36% (9)	50% (10)	52% (11)	36–68%
Setting the purpose	52% (13)	19% (4)	28% (7)	60% (12)	57% (12)	19–60%
Looking for patterns	24% (6)	14% (3)	4% (1)	30% (6)	0% (0)	0–30%
Think about my thinking	76% (19)	52% (11)	40% (10)	40% (8)	29% (6)	29–76%
Summarizing	20% (5)	43% (9)	8% (2)	35% (7)	14% (3)	8–43%
Number of students who responded =	(25)	(21)	(23)	(20)	(21)	

indicated if they had asked themselves particular questions during their work. Students were more likely to mark "Did I do what was expected?", "Am I focused on the assignment?", "Did I know everything I needed to know for the assignment?", and "Have I done my best work?" Marking these responses may have been self-serving in that students wanted to show that they were responsible and on task. Students were less likely to indicate that they asked themselves questions phrased in the progressive tense, indicating an action in progress. In order to ask themselves, "How well am I focusing on my assignment?", they would have had to engage in reflective thinking, a frame of mind with which they were less familiar. The same was true of "Am I understanding the assignment?" This question implied that as students worked, they checked their own thinking and comprehension of the task.

THINKING LOGS
Thinking Log entries documented changes from the beginning of the year to the end of school. Early in the year, student entries were perfunctory, such as "I am thinking about my science project" or "I am working on science." By the end of

the year, students showed signs of reflecting on their own learning. One student wrote, "How can I make my project better?" A student recognized confusion during math problem solving: "I don't understand this one bit." While writing a story, two students reflected, "This is confusing, I should ask Mr. T more about my paper" and "I am not sure what to write. I think I need some more details."

DISCUSSION AND IMPLICATIONS FOR FUTURE CLASSROOM RESEARCH

Based on written and observational data, it was clear that students needed help in developing a language for reflecting on their thinking while engaged in science activities. Students' default responses tended to emphasize task completion rather than learning. This implied that teacher modeling of how to talk about various types of problems that students encounter can lead to improved responses, as shown in the end-of-year Thinking Logs. A sample log entry is "I will describe my problem to someone else and see if that helps me come up with ideas." Students also needed descriptive language to express feeling confused about an idea or uncertain about what to do next. For example, Tomlinson modeled the following thought, "I am stuck and can't think of ways to test my science hypothesis. I need to find ways to generate ideas." When teachers provide students with appropriate stems to generate more reflective Thinking Log responses, students will be able to describe problems effectively and generate appropriate strategies to follow. This also supports two developmentally challenging aspects of metacognition: (1) being aware of learning problems and (2) identifying what to do. Tomlinson also plans to develop graphic organizers that explicitly emphasize the cognitive strategies incorporated into each science activity.

Students held a holistic view of each strategy. They did not know the meaning of some key terms such as *prior* or have a clear idea of the meaning of *knowledge*. Students had difficulty describing the nature of a pattern. Tomlinson will leverage his own new knowledge by thinking about the terms used to label the strategies. For example, metacognition could be labeled by a question, "What am I thinking?", prior knowledge could be labeled similarly as "What do I already know?", and pattern could be labeled, "What regularities do I see in size, shape, number, or color?"

In summary, strong parallels exist between how students learn and the processes involved in scientific inquiry and, by extension, reading comprehension. The complementary nature of reading comprehension and scientific inquiry formed the foundation for this research project on cognitive strategies. Knowledge is more than knowing discrete facts; knowledge must be transferable and applicable to new situations. By explicitly teaching and modeling cognitive strategies, teachers can provide more meaningful and essential learning experiences for their students.

LINKS TO THE NATIONAL SCIENCE EDUCATION STANDARDS

Strong parallels exist between how students learn and the processes involved in scientific inquiry, and by extension, reading comprehension. The complementary nature of reading comprehension and scientific inquiry formed the foundation for this research project on cognitive strategies. Content Standard A, K-4 (NRC 1996, p. 121) states that a result of activities in grades K–4 should be that students develop abilities to do, and understandings about, scientific inquiry. Among the abilities for doing inquiry are (a) asking questions about objects, organisms, and events and (b) planning and conducting investigations. These are sophisticated tasks for all learners. Scaffolding by Mr. T for determining importance in a passage of reading complements his work in scaffolding student questions and determining a purpose for an investigation. Both types of thinking require cognitive strategies for sorting important information from unimportant information and deciding on a purpose for action. One further example helps make the connection between thinking strategies for reading comprehension and those for doing scientific investigations. An understanding about inquiry targeted by the Standards is "Scientists develop explanations using observations (evidence) and what they already know about the world (scientific knowledge) (NRC 1996, p. 123). Mr. T guided students to apply what they already know to understand the text and use text information to construct summaries. From this line of thinking, students drew inferences based on information from the text. In science, students were guided in thinking about what the evidence means and how to summarize that meaning as it related to the original problem. In so doing, they needed to consider what they already knew and how this new information relates to that knowledge.

REFERENCES

Blumenfeld, P. C. 1992. The task and the teacher: Enhancing student thoughtfulness in science. In *Advances in research on teaching. Vol. 3: Planning and managing learning tasks and activities*, ed. J. Brophy. Greenwich, CN: JAI Press.

Bransford, J. D., A. L. Brown, and R. R. Cocking. 2000. *How people learn: Brain, mind, experience, and school*. Washington, DC: National Academy Press.

Brown, A. L., and J. C. Campione. 1990. Interactive learning environments and the teaching of science and mathematics. In *Toward a scientific practice of science education,* eds. M. Gardner, J. G. Greeno, F. Reif, A. G. Schoenfeld, A. DiSessa, and E. Stage. Hillsdale, NJ: Lawrence Erlbaum.

Brown, A. L., and J. C. Campione. 1998. Designing a community of young learners: Theoretical and practical lessons. In *How students learn: Reforming schools through learner-centered education,* eds. N. M. Lambert and B. L. McCombs. Washington, DC: American Psychological Association.

Cook, L. K., and R. E. Mayer. 1988. Teaching readers about the structure of scientific text. *Journal of Educational Psychology* 80: 448–456.

Dole, J. A., G. G. Duffy, L. R. Roehler, and P. D. Pearson. 1991. Moving from the old to the new: Research on reading comprehension instruction. *Review of Educational Research* 61: 239–264.

Flick, L. B. 2005. Being an elementary science teacher educator. In *Contemporary issues in elementary science teacher education*, ed. K. Appleton, 15–29. Hillsdale, NJ: Lawrence Erlbaum.

Flick, L., and M. Tomlinson. 2001. The role of reading in teaching scientific inquiry. *The Oregon Science Teacher* 42: 9–12.

McNeil, J. D. 1987. *Reading comprehension: New directions for classroom practice.* Glenview, IL: Scott, Foresman.

National Research Council (NRC). 1996. *National science education standards.* Washington, DC: National Academy Press.

Wittrock, M. C. 1998. Cognition and subject matter learning. In *How students learn: Reforming schools through learner-centered education,* eds. N. M. Lambert and B. L. McCombs. Washington, DC: American Psychological Association.

RESOURCES

Billmeyer, R. and M. L. Barton. 1998. *Teaching reading in the content areas: If not me, then who?* Aurora, CO: McREL.

Ciborowski, J. 1998. *Textbooks and the students who can't read them.* Brookline, MA: Brookline Books.

Farstrup, A. E. and S. J. Samuels. 2002. *What research has to say about reading instruction.* Newark, DE: International Reading Association.

Rhodes, L. K. 1993. *Literacy assessment: A handbook of instruments.* Portsmouth, NH: Heinemann.

Scala, M. C. 2001. *Working together: Reading and writing in inclusive classrooms.* Newark, DE: International Reading Association.

Schoenbach, R., C. Greenleaf, C. Cziko, and L. Hurwitz. 1999. *Reading for understanding: A guide to improving reading in middle and high school classrooms.* San Francisco, CA: Jossey-Bass.

Scott, J. 1993. *Science and language links: Classroom implications.* Portsmouth, NH: Heinemann.

Snow, C. E., S. Burns, and P. Griffin. 1998. *Preventing reading difficulties in young children.* Washington, DC: National Academy Press.

Tierney, R. J., J. E. Readence, and E. K. Dishner. 1995. *Reading strategies and practices.* Boston: Allyn and Bacon.

Tovani, C. 2001. *I read it, but I don't get it.* Portland, ME: Stenhouse.

Weaver, C. 1994. *Reading process and practice.* Portsmouth, NH: Heinemann.

Wollman-Bonilla, J. 1991. *Response journals.* New York: Scholastic.

AUTHOR AFFLIATIONS

Lawrence B. Flick is a professor of science education and chair of the Department of Science and Mathematics Education in the College of Science at Oregon State University. He taught middle level and elementary science for 13 years and elementary, middle school, and high school teacher education for 25 years. He holds a PhD in science education, a master's in education, and BS in electrical engineering. Flick's work examines the nature of instruction and the role of formative assessment in developing student skills in, and knowledge of, scientific inquiry.

Michael Tomlinson has taught grades 1–6 for 22 years and currently teaches at James Templeton Elementary School. He has been an active member of the Oregon Science Teachers Association and the Science Teacher-Leader Cadre for the Oregon Department of Education. Receiving a Christa McAuliffe Fellowship, Tomlinson used the funds to further his interest in integrating instruction in reading comprehension with higher-level thinking in science. He holds a BA and master's degree in elementary education.

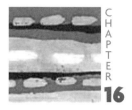

revised views of classroom assessment

Mistilina Sato, Vicki Baker, Elaine Fong, Joni Gilbertson, Tracey Liebig, and Neil Schwartzfarb

ASSESSING CLASS PROJECTS: A CONVERSATION TWO YEARS AGO

The following conversation demonstrates the typical assessment practices of middle school science teachers from the New Haven School District in Union City, California, two years before they participated in the Classroom Assessment Project to Improve Teaching and Learning (CAPITAL).

Joni: *Piles and piles of paper.*

Tracey: *Just look in a teacher's car sometime.*

Neil: *That will give you the big picture of everything a teacher collects from students.*

Vicki: *It's Friday afternoon and someone asks you what you are doing this weekend. I have 195 element projects to correct. I sit and read them all night. Some are really neat. Some are awful. I read each one of them carefully and mark them thoroughly.*

Joni: *And by 11:30, your standards are sliding . . . .*

Vicki: *I have to decide, should I give it an A- or a B+. Eventually, I start not caring if they missed an important point in the assignment because I know there will be no going back.*

Neil: *Then when you hand them back, Johnny drops his project in the garbage can without even looking at your comments . . . .*

Vicki: *Or someone asks why she got a B+ and, for the life of you, you're not sure why that project got a B+.*

Tracey: *It could be because you graded it at 7:00 and not at 11:00.*

> *Vicki: And I go around thinking I'm a terrific teacher because I assigned a great project and I get the graded projects back to kids with quick turnaround.*

BACKGROUND

CAPITAL, a collaborative National Science Foundation-funded research initiative between Stanford University and nearby school districts examined classroom-based assessment in science. University staff met regularly with 30 teachers individually and in small groups to discuss their current and changing assessment practices. The teachers shared ideas with one another, and the university staff introduced research findings and ideas from other teachers into the conversations. During CAPITAL meetings, teachers delved into the underlying reasons guiding their assessment practices.

The CAPITAL staff and teachers believed that change in classrooms must be grounded in teachers' practical reasoning, or reasoning directed toward taking principled action in their own classrooms (Atkin 1992). This starting point rejected the dichotomy often presented between knowledge and action, and between research and its application. People with practical responsibilities necessarily use what they know and believe to make decisions in their own settings (Schwab 1969).

CAPITAL also drew on the research literature on classroom formative assessment—assessment aimed not only at evaluating student achievement, but also at supporting student learning. Black and Wiliam (1998a, 1998b) reviewed empirical studies of classroom assessment and concluded that research supports the argument that formative assessment practices improve student achievement and learning. CAPITAL drew upon formative assessment approaches such as the nature of feedback (Butler 1987; Darling-Hammond, Ancess, and Falk 1995; Davis 1996; Tunstall and Gipps 1996), effective forms of goal setting (Sadler 1989), and self- and peer assessment (Rudd and Gunstone 1993; Sadler 1989; White and Frederiksen 1998).

The five teachers from the New Haven Unified School District who participated in CAPITAL tried new approaches and techniques in their classrooms, meeting monthly with one another and university researchers to examine and discuss their evolving practices and classroom innovations. While these monthly meetings were the primary setting for interactions, the teachers were in regular communication with one another at their school sites and through e-mail. Each teacher tried different strategies and shared a desire to make the assessment processes in his or her classroom more focused on learning. Given that each teacher had contact with 160–180 students per day in heterogeneously grouped and culturally diverse classrooms, all the teachers also wanted to develop means of assessment that were not overly time intensive for written feedback and evaluation.

The monthly CAPITAL meetings were videotaped, with teachers periodically writing reflections about their work. The primary sources of data reported here were transcribed conversations among the teachers and individual cases written by the teachers in June 2002.

FINDINGS

For the CAPITAL teachers, a shift from assessment *of* learning (as depicted in the opening conversation) to assessment *for* learning occurred. Teachers came to zero in on questions such as the following:

* Why did I assign this project? What were the learning purposes?
* What information was I getting about what the students had learned from the project and from my mode of assessment?
* What was worth my time and the students' time when assessing student's work?
* How can I better use my time to help students learn while maintaining efficiency in my assessment practices?

Monthly conversations among the teachers frequently fostered innovations in assessing projects. CAPITAL researchers introduced the teachers to the findings from Butler's study (1987)—students who were given grades alone or grades with comments showed little to no learning gains, while students given comments without grades showed the most growth. This triggered a discussion about assigning grades to projects. Some teachers decided to assess project assignments using an *Accepted/Not Accepted* framework, rather than points or letter grades. *Accepted* meant that the student achieved the standards established for the project. *Not Accepted* meant that the assignment would need to be revised until it was judged to be *Accepted*. There was no middle ground—either the standards were met or they were not met. Following are examples of how this assessment process evolved for three of the five teachers from New Haven Unified School District.

VICKI'S ASSESSMENT PROCESS

Projects constitute about 30% of my students' grades. I decided to make a project where Accepted would be an A+; a Not Accepted project had to be revised. Once revised, it too would receive the same A+. When I first presented this plan to my students, about half of them thought the idea was exciting. The other half either didn't quite understand or didn't believe me. They asked questions such as, "What if it's not accepted the second time?" Then I would ask them to revise it again until it met the standards. There were no deductions for the revision process because, in the end, I knew the student had met the standards I expected—it just took some more feedback and opportunity to revise. At first, it was a big risk because I didn't know how long revisions would take or how long I would have to extend the timeline on projects. It worked out, though, and I have had an increase in how many projects get turned in, which is worth the extra time for the revision process. I expected some parents to think this method might be unfair to their children, especially students who put extra work into their projects, but I heard from no one.

My first project was one that Tracey had developed. Students picked any item that

moved a distance and calculated its average speed in 10 trials. I created a list of everything needed to get Accepted on the assignment and distributed it to the students. I told
the students to return the list with their completed assignments.

I had no idea what to expect when I took the projects home to assess. I was pleased
to see that I still got beautiful, high-quality work from the high achievers. Though
grading was much quicker than usual, I was a little disappointed in the results. Only
about one-third of the students received Accepted marks. As I handed back the Not
Accepted papers, I worried about how seeing Not Accepted might affect my students.
I expected them to take their papers home and mull over what they did wrong. What
happened instead was better. They took them back to their desks and helped each other
and asked me for help. Within a few minutes, students returned revised projects. My
favorite moments came when students who probably had never received an A+ asked
me, "What is my grade on this?" I had to repeat "A+" several times to some students.
Some asked me to write it down for their parents. Several stood behind me while I en-
tered A+ into my computer.

I think that if something is worth learning in my class, I should make sure that
everyone is given a chance to learn it and revisit it if he or she didn't doesn't learn it
the first time around. It is faster to grade projects using this method. The second time
doesn't take long because I circle what is required on the list of expectations. I don't
think students are bothered by the fact that everyone gets the highest grade. The students know that their work is being graded against a set of expectations they know
about in advance so the assessment is not as arbitrary as my old way of assigning
grades was. The students also seem to have their own internal standards, and so I
continue to see a wide range of quality in how the final projects appear. I make sure to
focus on the substance of the project and not on the aesthetics.

Tracey and Joni used Accepted/Not Accepted with a lot of peer assessment. I look
forward to trying some of their ideas. Neil and I discuss and disagree on what to do
with late projects; I still deduct points for turning in assignments late. I think I'm coming around to his way of thinking (accept them for the same grades). I need to continue
to share what worked and didn't with my CAPITAL group.

TRACEY'S ASSESSMENT PROCESS

As I approach my classroom, students are gathered in the hall analyzing the posted
work. "That fish is a bottom dweller but its eggs are floating. What's up with that?"
critiqued Nicole. Sadiq responded, "I asked the same question, and Heather wrote
about the reproductive adaptation on the back of the paper. She wrote that the babies
would hatch away from the typical predators of the parent fish." Sarah said, "I was
thinking the same thing," then pointed to another poster and asked, "What do you
think about these colorful vertical stripes?"

This conversation occurred nearly three months after the lesson on adaptations in
which the students created fish with adaptations and explained the reasons behind
their choices. I had done this same lesson over the years and this long-lasting interest

was new. In the past, student conversations consisted of a brief look at the fish poster display and short comments about a particularly attractive fish. Students often threw their work away after reading my assigned grade, ignoring my written comments or questions. What was different?

Through collaboration with my CAPITAL group, I used the rubric format designed by Elaine that included a list of criteria for the project; the revision process shared by Joni, in which students had time in class to revise their work after a peer assessment but before submitting it to me; the Accepted/Not Accepted grading system developed by Vicki [described above]; and the analysis of example work that Neil found successful. The adaptation project came after several other projects in which I had used some of these assessment ideas and had gotten positive results from the students.

For this project, I modeled the type of feedback that would guide peers toward an acceptable product. I also had the students write a note to their peers based on the project requirement list; they were required to include questions to guide the revision process. The project requirements said, "Describe reproductive adaptations and the reason behind the success," and a typical peer question was, "I see that your fish buries its eggs. Why is this a good survival strategy?"

Students used the questions to guide the project revisions. Two special needs students told me with confidence that their fish was going to be returned acceptably complete and that they wouldn't need my help since their peers had guided them. Previously, both had the impression that they needed my support toward acceptable completion.

Why was this different? I think students were becoming accustomed to the fairness of the revision toward Accepted/Not Accepted criteria and the validity of their opinion in the critique process. I think they found it enjoyable enough to continue to act in that role for months afterward. I took ideas away from the collaborative CAPITAL process and made the practices my own. As I looked back at the work considered exemplary I had saved over the years, work I had spent so much time evaluating and grading, it wasn't to the high standard I got this year from a much broader group of students with this new system of assessment.

JONI'S ASSESSMENT PROCESS

I discovered the power of peer assessment and began experimenting with it in various forms in my classroom. One example is the cell project in which students created a three-dimensional model of a plant or animal cell. I created a checklist of criteria with three parts—one for the student, one for a peer, and one for me. The purpose of the two student rubrics was to give the creator an opportunity to revise the model before I graded it. At first, I was skeptical about using peer assessment. Would students actually benefit from using the class time in this way? Would students automatically check all the yeses on the rubric to be nice to a friend while peer reviewing?

First, I asked the students to work in groups to assess models from the previous year. They all wanted to know how I would grade the models. Some students seemed stunned that every eye-catching model did not have correct components. When their models were complete, I asked the students to use the checklist to assess their own models and revise any parts that were not acceptable. Then the students sat with a peer and assessed each other's models. They then revised their models without penalty before I assessed them.

I was nervous about what might occur but let them get to work. I was amazed. Everyone was eager to work. Based on the conversations I heard, students were taking the responsibility of assessment seriously. They referenced their textbook and notes to make sure things were correct on the models. More than that, learning was taking place through the assessment process as each student had to be knowledgeable about cell structure in order to assess the models. They were helping each other.

I eagerly shared my experiences with the CAPITAL group. It was enlightening to analyze the lesson and discuss my concerns. Much of the discussion centered on fairness. Was it fair that students who fixed their model could get the same grade as someone who did it perfectly the first time? Was it fair not to give extra points for prettiness? Was it fair that hastily put together models could be fixed up in class? Would students stop producing quality work if they could get the same grade with less effort? We came up with many ways of looking at these questions. We concluded that if a student understood the concept, it didn't really matter in the end if it took them three attempts.

I see this as an evolving process in my classroom. I like the role I have when I am working with students to revise or redo. I like the emphasis it places on learning instead of on grading the project. I also feel that one of the most valuable things is that it places the responsibility for learning onto the student. By determining in advance what my goal is and giving every student the opportunity to achieve that goal, I have created a more equitable classroom. I discovered that those students who were driven to do extraordinary work remained driven. I also realized that some of the artistic students didn't demonstrate understanding of concepts even if their projects appeared beautifully constructed. Focusing on the content-based criteria helps the students create more meaningful work.

ASSESSING CLASS PROJECTS: A CONVERSATION TODAY

The following conversation reveals the changes experienced by the teachers in the first conversation (pp. 197–198) since participating in CAPITAL.

Joni: I will never forget the feeling of excitement I had…that initial sense of awe I had when I looked around the classroom and saw kids focused, talking about the cells, helping each other fix their projects to make them acceptable. It was like a magic door opened up.

Vicki: The lightbulb for me from your story was letting the students fix their projects in class. At first, I thought, "Is that really fair?"

Joni: It really leveled the playing field for everyone. Parents weren't doing the projects. Kids who didn't have craft resources at home could work on their projects. All the kids had just as much chance to get an A as those who had all *the craft materials at home.*

Vicki: *And you realized that some kids who have access to the craft materials may not actually get the ideas.*

Joni: *The second year I did the cell project, kids could use the rubric to assess the previous year's projects so they better understand the expectations. That's a layer I got from Vicki when she helped the kids evaluate a previous year's project with the same rubric that would be used for evaluating their work.*

Tracey: *The first year I tried this technique, I used the exemplary examples and now I use a range of examples so the kids get a better understanding of what the rubrics mean.*

Neil: *The difference for me is that I now get projects from everyone, not just the students who would turn in pretty ones.*

Joni: *My role as the teacher changed. I became an adviser. I gained a lot more information about the students from interacting with them.*

Vicki: *The process looks the same on the surface, but a closer look shows you that the students are more invested in the work they are producing.*

Tracey: *I now have rubrics and checklists and examples of student work from previous years. I use a peer review process that allows the students to revise their work prior to turning it in to me so by the time I finally grade it, it's the students' best work.*

Neil: *I have fewer projects to grade all in one sitting. The kids hand them in intermittently.*

Tracey: *I have greater clarity in the grading process. The rubrics and the peer review process nail down the expectations. The expectations are clear at 1:00, they are the same at 11:00 that night, and it's clear when you explain it to parents.*

Vicki: *I don't get as frustrated. I used to feel like I told the kids over and over what to do. Now, I know the expectations are clearer to them so when I see a project that misses the mark, I now ask myself, "Why didn't they get it?" I'm self-evaluating more.*

Joni: *I'm really appreciating being clearer in my expectations.*

Vicki: *I find that I have to do a lot more work up front when I assign a project, but it pays off.*

Joni: *Helping kids revise and redo while they were doing their projects gave me the opportunity to provide constant feedback while they were working. I think students learn more when they get that immediate feedback rather than waiting for me to grade their projects and hand them back three weeks after they were completed.*

Joni, Vicki, Tracy, and Neil began their discussions around a practical desire to develop means of assessments that would not result in overwhelming paperwork but that would allow their students opportunities to learn through assessment processes. Developing and implementing the *Accepted/Not Accepted* framework moved the teachers closer to this goal. New learning opportunities for students through continuing the learning dialogue between teacher and student and among students occurred. Understanding concepts and meeting standards were nearly universally motivating for students, whereas student motivation for grades varied. A learning environment centered on students being allowed to revise their work led to collaborative peer relationships in some classrooms. In many cases, students completed projects and met deadlines more promptly when given the option of resubmitting revised work after receiving feedback from the teacher or from peers. Revision became such a strong theme that even teachers who did not fully adopt the *Accepted/Not Accepted* grading framework incorporated revision processes into their classroom assessment.

The teachers viewed the *Accepted/Not Accepted* assessment framework as addressing issues of fairness and equity. Initially, it seemed unfair to allow students who did not produce nicely crafted work to earn the same grade as students whose projects were more aesthetically pleasing. It became clear, however, through the teachers' discussions, that they valued conceptual understanding more than the craftsmanship of the projects. Using the revision process provided opportunities for teachers to work with individual students on particular areas. Some teachers talked about identifying student misconceptions and addressing them on the spot, rather than waiting until the project was graded.

CONCLUSIONS

The teachers described themselves as moving away from the role of teacher as giver of grades to teacher as conductor of learning. They increased their interactions with students during class time and described seeing student growth where there had previously been frustration. When the standards for assessment became clearer to the teachers, they felt that they could better assist students toward those learning goals. Teachers felt more confident that the students' grades represented concepts attained, not just work completed.

These stories from classroom experiences highlight how the interactions among the CAPITAL group helped shape the direction that the teachers individually took to develop new assessment practices. In the past, teachers had few opportunities for sustained conversations focused on practice. CAPITAL was designed to provide collaborative opportunities for groups of teachers to share ideas and analyze classroom practice. As Vicki stated,

It is through our collaboration that we begin to understand what an art teaching is. One of us will come to the group with an exciting idea. . . . We will sit for long periods of time discussing the idea, down to the most minute details. Then we take the ideas back to our own classrooms. If we decide to use it, we'll change it to make it work for us. We then share what we did and how it worked for our particular students and within our teaching philosophy. Another change might be made in another classroom. Sharing ideas definitely does not mean that we all are doing it the same way.

The CAPITAL teachers did not opt for cookie-cutter innovations. They sparked ideas, critiqued their own techniques, offered suggestions, and reflected on how innovations in assessment worked for their students.

ACKNOWLEDGMENT

The Classroom Assessment Project to Improve Teaching and Learning (CAPITAL), a collaborative research initiative between Stanford University and nearby school districts, was funded by the National Science Foundation (Grant # REC-9909370).

LINKS TO THE NATIONAL SCIENCE EDUCATION STANDARDS

The assessment focus of the Classroom Assessment Project to Improve Teaching and Learning (CAPITAL) centered on the interactions between teachers and students and among students. Several standards from the National Science Education Standards (NRC 1996) were relevant to the work of this project and the teachers' classroom assessment explorations (Teaching Standard A and C). While participating in CAPITAL, teachers shifted from using assessment for purely evaluative purposes to using assessment strategies, such as revision of work, peer assessment, and self-assessment, for learning opportunities. The standards for Assessment in Science Education outlined the purposes of assessments conducted by classroom teachers (pp. 87–89). The work of the CAPITAL teachers reflected their efforts to "develop self-directed learners" (p. 88) through processes of self- and peer assessment. Specifically, they developed criteria and expectations for projects to be completed by the students. By making these expectations available to students during class projects, students evaluated their work as it progressed and, in some classrooms, students received feedback from peers about their work.

REFERENCES

Atkin, J. M. 1992. Teaching as research: An essay. *Teaching and Teacher Education* 8 (4): 381–390.

Black, P. J., and D. Wiliam. 1998a. Assessment and classroom learning. *Assessment in Education* 5 (1): 7–74.

Black, P. J., and D. Wiliam. 1998b. Inside the black box: Raising standards through classroom assessment. *Phi Delta Kappan* 80 (2): 139–148.

Butler, R. 1987. Task-involving and ego-involving properties of evaluation: Effects of different feedback conditions on motivational perceptions, interest and performance *Journal of Educational Psychology* 79 (4): 474–482.

Darling-Hammond, L., J. Ancess, and B. Falk. 1995. *Authentic assessment in action: Studies of schools and students at work*. New York: Teachers College Press.

Davis, B. 1996. Listening for differences: An evolving conception of mathematics teaching. *Journal for Research in Mathematics Education* 28 (3): 355–376.

National Research Council (NRC). 1996. *National science education standards*. Washington, DC: National Academy Press.

Rudd, T. J., and R. F. Gunstone. 1993. Developing self-assessment skills in grade 3 science and technology: The importance of longitudinal studies of learning. Paper presented at the Annual Meeting of the National

Association for Research in Science Teaching, Atlanta, GA (April).

Sadler, R. 1989. Formative assessment and the design of instructional systems. *Instructional Science* 18: 119–144.

Schwab, J. J. 1969. The practical: A language for curriculum. *School Review* (Nov.): 1–23.

Tunstall, P., and C. Gipps. 1996. Teacher feedback to young children in formative assessment: A typology. *British Educational Research Journal* 22 (4): 389–404.

White, B. Y., and J. R. Frederiksen. 1998. Inquiry, modeling and metacognition: Making science accessible to all students. *Cognition and Instruction* 16 (1): 3–118.

Resources

Black, P. J., and D. Wiliam. 1998. Inside the black box: Raising standards through classroom assessment. *Phi Delta Kappan* 80 (2): 139–148. Available at *www.pdkintl.org/kappan/kbla9810.htm*.

Stiggins, R. J. 2002. Assessment crisis: The absence of assessment FOR learning. *Phi Delta Kappan* 83 (10). Available at *www.pdkintl.org/kappan/k0206sti.htm*.

AUTHOR AFFILIATIONS

Mistilina Sato, an assistant professor of teacher development and science education at the University of Minnesota, Twin Cities, was a research assistant on the Classroom Assessment Project to Improve Teaching and Learning. She taught middle school science before receiving a PhD in science education from Stanford University.

Vicki Baker, Elaine Fong, Joni Gilbertson, Tracey Liebig, and **Neil Schwartzfarb** are middle school science teachers in the New Haven Unified School District in Union City, California. They are professionally active in improving science education in their schools, district, and state.

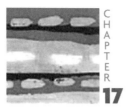

moving beyond grades and scores: reconsidering assessment feedback

Caleb Cheung, Anthony Cody, Irene Hirota, Joe Rubin, and Jennifer Slavin
with Janet Coffey and Savitha Moorthy

INTRODUCTION

The comments that you write need to give kids something to do. Rather than writing a question in the margin, it makes more sense to give them something to do, to make it an action thing, which they can read and feel, "this is what I need to do."

And I think my students have started to pick up on...I mean, there'll be times when I have to read stuff in the morning and give really quick feedback. Sometimes all I do is circle stuff or put a line through it, and my students are like, "I don't understand when you write 'no.'" They are aware that they don't know what they need to do, that they have to ask me.

It's a really different dialogue than when I'm reading something and giving thoughtful feedback, that they can read and do something with rather than say, "This part was wrong." And they recognize the difference now— they know when they get feedback that they don't know what to do with.

(Jen; February 2002)

BACKGROUND AND RESEARCH CONTEXT

The Classroom Assessment Project to Improve Teaching and Learning (CAPITAL) examined the intersection of teacher change, teacher beliefs, and assessment in the science classroom. CAPITAL drew on findings about assessment and teacher change. Black and Wiliam (1998a) examined numerous studies that focused on aspects of formative assessment, such as questioning, feedback, self- and peer assessment, and assessment criteria. One central feature of formative assessment was the feedback loop between teaching and learning (Ramprasand 1983). Information gleaned from formative assessments should link back into the learning cycle. Feedback also informs teachers about modifications they need to make to promote more effective learning.

This study drew on several schools of thought regarding change in the classroom, especially conceptions of teachers' practical reasoning and how it is developed through experience. The practical reasoning of teachers was directed toward taking principled action in their own classrooms (Atkin 1992). This way of thinking assumed that people with practical responsibilities necessarily use what they know and believe to make decisions in highly contextualized settings (Schwab 1969). During these processes, teachers developed repertoires of actions that were shaped both by standards and by knowledge gleaned in practice (Wenger 1998).

Research literature suggested that few classroom strategies were as effective as formative assessment in improving learning for all students (Black and Wiliam 1998a). Teachers' underlying views (about teaching, learning, students, subject matter) influenced instructional changes (Atkin 1992). The CAPITAL staff believed that a concerted effort to help teachers incorporate these views in an integrated fashion *into their own ways of working with students* would yield important advances in the quality of assessment, and subsequently, student learning.

Project staff (researchers at Stanford University) worked with San Francisco Bay Area teachers who, though they believed that high-quality assessment was a central feature of good teaching, felt that their existing practices fell short of their own goals. From the beginning, the staff did not tell teachers what to do, nor did the staff or the teachers subscribe to a one-size-fits-all model for instructional change. The teachers experimented with new approaches and techniques in their classrooms and, in collaboration with the researchers, kept careful records, and collected students' work. They also met regularly with other teachers and the university researchers to examine and discuss their evolving practices and classroom innovations.

Teachers and researchers together developed narratives of the teachers' assessment practices that helped them (1) better understand the frameworks within which the teachers operated and (2) encouraged teachers' self-examination of these frameworks. While researchers learned about individual teacher practices and the reasons for those practices through the teachers' narratives, the researchers also developed a more sophisticated understanding of classroom assessment and teacher change as related to assessment. Researchers compiled detailed notes of class observations, had reflective conversations with the teachers, and analyzed videotapes of classroom and teacher group discussions. In addition, researchers and teachers collected and analyzed other data from the classroom, including student work samples and associated teacher commentaries.

FINDINGS

Over time, the CAPITAL teachers broadened their notions of assessment, moving beyond grades and scores to include feedback. This shift led to student learning as *the* integral dimension of any assessment activity. As the CAPITAL teachers examined their assessment practices, the type, quality, and content of feedback emerged as an interest and focus for discussion. For the teachers, coming to understand the meaning of "effective" feedback and how it operated in their classrooms was not a straightforward journey. The next section contains reflections from five teachers on their assessment practices—specifically, the act of providing useful and constructive feedback to facilitate student learning.

JEN'S PERSPECTIVE (USING PEER REVIEW WITH STUDENTS)

When I listen to my students share their understanding in class, I often feel they understand the concepts. However, when they write short-answer tests I am frequently not convinced that they understand the content as well as I thought they did, or would like them to. My students' writing raised a very important question: Do they not know the science well enough to clearly explain it? Or do they not know the mechanics of good, clear writing? Since poorly written science answers didn't help me assess their understanding of science, helping students improve their writing could not just be left to English teachers.

If I was going to hold my students accountable for their writing, however, I needed to help them learn how to write better. So, I began using writing lessons as one way to teach science content. This choice to focus on writing involved some compromises. I reluctantly included more time to revise drafts in class and less time doing hands-on activities and worried that the class would be less engaging for my students, and for me. I also worried that I lacked the skills to teach writing since it was outside my area of expertise.

Initially, my students had trouble writing information in a logical order, making it very difficult for me to determine how correctly they were thinking about the science. My students and I discussed how to write strong sentences and how to organize them into solid paragraphs. I read drafts of their writing and offered feedback, and students revised their work. Writing meaningful feedback on students' first drafts was a very laborious process. I had to write feedback about the mechanics of writing as well as the science content; it was unclear whether feedback on the science, on their writing, or both would be the most beneficial to their learning.

I wrote planning outlines for questions to scaffold the paragraphs so students could organize their writing better. This added structure and made it easier for me to see what my students understood about the content. As a result, my feedback became more directed at their understanding or misunderstanding of the science.

My students were conscientious about my feedback. They diligently "fixed" their paragraphs and changed things I suggested. However, I began to wonder whether they were really thinking critically about their revisions and about my feedback. I felt my students often just accepted what I said as true: They revised their paragraphs without revising their thinking, and without learning.

Since we use many group-work activities in my classroom, I turned to peer review and had students critique each other's work against the assignment's rubric, hoping they would evaluate their peers' feedback more critically than they had my feedback. There were times when students' comments were irrelevant or even inaccurate; however, this pushed the student who received the feedback to think more critically to evaluate it.

An unexpected bonus of peer review was that students began to challenge their own understanding as they reviewed a classmate's work. They had to decide whether their classmate's work was clearly written, complete, and scientifically accurate, forcing them to clarify these ideas for themselves. Students asked me questions that indicated they were pushing their peers to explain concepts completely and as a result had to decide what they thought a complete explanation was—for example, "He's written that there are a lot of blood vessels around the brain, but he hasn't said why. It's pretty important why there are a lot of blood vessels around the brain, isn't it?" In other cases, students asked questions that told me they were thinking about the accuracy of the content on their peers' papers and comparing that with what they held to be accurate—for example, "Would there still be dopamine inside the dendrites of the message receiver even if there's cocaine in the synapse? Does any dopamine get into the next neuron to keep the signal going?" As they reviewed their peers' work, students tried to decipher what was important and what was not as important, what level of detail was required and what was excessive.

When going back to their own work, students were able to revise their answers because their thinking had been challenged not only because of the feedback I had written but also by the act of giving and receiving feedback on a peer's work. Over time, students' writing became clearer and more complete, allowing me to see more clearly what my students understood about the science. Only at this point could I use their writing to accurately gauge their understanding.

ANTHONY'S PERSPECTIVE (MODELING THE PEER-REVIEW PROCESS)

For my sixth-grade Earth science classes, my goal is to create engaging, open-ended challenges that capture students' imaginations and deepen their knowledge. I also want students to be able to communicate their understanding with others. The study of the planets in our solar system is primarily a question of researching and reporting. I wanted to avoid the dry lists of facts that research projects often elicit, so I challenged my students to create a project incorporating factual information. They could create an illustrated cartoon, a travel brochure, or a science fiction story. I gave the students reference materials to use in class, and class time for research.

After a week of work, students' drafts, in most cases, were in need of improvement. I then decided to introduce peer assessment into the process. I believed it would cause students to understand the assignment better, since they would be struggling to assess the work of another. I also saw this as a chance to increase the amount of feedback students would get. I created a review sheet for students to use as they looked at their peer's work. The review sheet called for a numerical score, comments on the strengths of the work, and suggestions for improvement.

I also distributed a detailed rubric that described the elements a project would need to score a 4 on a 4-point scale:

* A list of facts about the planet, including length of day and year, gravity, distance from the Sun, atmosphere, surface features (what is it made of?), description of moons (if any), other interesting details.
* Vivid descriptions and an engaging plot (if the project is a story).
* Excellent drawings that illustrate a lively story (if the project is a cartoon).
* Excellent illustrations and descriptions of exciting, imaginative activities that could actually be done on the planet (if the project is a travel brochure).

A project that rates a 4 is strong on science and fiction both!

I found the students quite capable of using the list of facts about the planet. However, they found it difficult to evaluate the "vivid descriptions and an engaging plot." That night I wrote a short science fiction story of my own. I tried to follow the rubric, weaving as much detail into my narrative as possible. I labeled it "first draft" and did not even finish it—it was only one-page long.

The next day, we began our lesson by reading my draft and then compared it to the rubric. How many of the facts were covered? Were the descriptions vivid? What made them so? What was missing from the plot? How could the story be continued and extended? The students had lots of ideas and seemed to actually be tapping into knowledge from their other experiences in writing as the discussion unfolded. They were not at all shy in criticizing my story, and I made it clear that it did not hurt my feelings; after all, the goal was to produce a stronger story, so their ideas were helpful to me.

This ended up being an exercise in modeling on two levels. My draft story provided a model for their own stories. As we critiqued my work, they could see how I used descriptive language and characters to engage the reader as well as to communicate information. We likewise modeled the peer review process, and I demonstrated that one could put forward work and take criticism without feeling put down or defensive. I also was able to coach them to give their comments in a constructive way, with specific suggestions as to what could be improved, along with praise for what was strong in the piece.

When we moved into the next phase, having students review one another's work, I was impressed by the intensity of a number of the conversations. Students were reminded that they were writing for an audience that included their peers, rather than simply their teacher. They responded to the drafts with concrete suggestions, instead of just saying, "It was good." A number of draft stories had featured paragraphs of facts, baldly stated. Students pointed out how these facts could be woven into the story. Some of the characters were no more than a name, so students gave suggestions to one another to describe the characters and make them come to life. Revision followed these *exchanges, and many of the projects were significantly improved as a result.*

CALEB'S PERSPECTIVE (GRADE REPORTING AS A TOOL)

During my first year of teaching I received a lot of advice about organizing my class-room to help my students succeed. Among the advice was to post grades regularly to hold students accountable for their academic performance. This made sense to me be-cause that technique had helped me when I was in school. Over the past six years, grade posting has become a regular part of my teaching routine. I started by creating a simple computer spreadsheet to show missing assignments. Then slowly it evolved into a collection of student records and formulas that calculate totals, percentages, and letter grades. Once a week, I posted a version of this spreadsheet with student identi-fication numbers in place of names to keep the grades confidential. I quickly realized how efficient a computerized grade book could be. I could easily calculate and adjust individual and class grades as needed.

Besides the amount of time I saved as a teacher, benefits to students also seemed significant. Students who were normally oblivious to their grades at the end of the report-card period became more aware of how their daily work resulted in the grades they earned. There were no secret grade formulas. I followed the standard percentages for grade assignments as set by the school. I did not play favorites. Everyone was treat-ed in exactly the same way except for students with special needs. When surveyed, all my students said they considered grade posting helpful. The students were never able to make excuses such as "I didn't know my grade" or "I don't know what I'm missing."

The list of completed and uncompleted assignments provided a concrete way for my students to take control of their grades. Students who were disorganized and had a difficult time keeping track of their work now had a clear picture of where they stood. There was increased motivation to complete and turn in assignments. Most of all, I thought I was providing weekly feedback about their academic achievement and behav-ior to motivate each student to improve. In line with my district's directive to improve grades and test scores, I was giving them a tool for success.

Over time, I realized there were a number of problems with posting grades. A col-league pointed out to me that grades often became the center of attention for my stu-dents and me. For example, when grades were posted at the beginning of the week, the attention of the whole class shifted solely to the grades. During the week, I found my-self frequently reminding students to check their grade and encouraging them to keep up with their work. Some students wanted to check their grade at every opportunity, creating constant anxiety about or obsession with grades. My school's emphasis on good grades further exacerbated the issue. Students were bombarded with the message that high grades were the main purpose of their education. I thought I was setting high expectations by providing accountability. Instead, the grades often took the focus of the class away from learning.

While most students used the grade postings to maintain their work, others bragged about As or stopped trying when they consistently received Fs. Discouraged, some of these students gave up. A student even told me his low grade "means the end of the world. I know I need to do a bundle of work, but now it's too late." Some students settled for a C or a B as "good enough" and saw no need to try harder. Even though I kept student numbers confidential, there was grade comparison and teasing among some students. I tried my best to address these issues by talking about them with the class, but students still labeled themselves or others based on the grade they received. The letter grades in and of themselves carried meaning outside of their academic accomplishments. For students, an A equated to excellent, good, or "knowing everything." An F meant you were "bad," "did not try at all," or "did not know anything," regardless of how much a student may have improved. Some A students received high grades because they were really good at completing assignments, not necessarily because they learned the material. I realized grades did not accurately reflect the learning accomplished by the students.

To counter some of these tendencies, I increasingly accepted late work and makeup work as a way for students to improve their grades. In their minds, completion became more important than learning. This led to some students copying and cheating to receive high grades at all costs. Most importantly, the feedback these grade postings provided were limited to completion. Areas for improvement and mastery of a skill or subject were not communicated in the grades.

I now feel that grade reporting is just a tool. Like many other techniques of assessment, it has its limitations. As a teacher in an urban, low-performing school, I wonder if grade posting has done more harm than good. Although grades are unavoidable in my school culture, are there alternatives I can provide in my classroom to motivate students to learn? What can I do to de-emphasize grades and emphasize learning? Is there a way to supplement or modify grade posting to present a more complete picture of a student's learning progress? How can I give appropriate feedback and teach 160 students a day?

JOE'S PERSPECTIVE (USING RUBRICS AND HIGH EXPECTATIONS)

After reading the article entitled "Inside the Black Box: Raising Standards Through Formative Assessment" (Black and Wiliam 1998b), I was shaken. The authors articulated the influence on learning of traditional assessment techniques—those that evaluate students with a letter grade but do not provide them with feedback or opportunities to revise their work and their understanding of concepts. I was the poster boy for misguided assessment techniques that put the focus on grades instead of learning.

In trying to make assessment an opportunity for student learning, I now give frequent shorter tests instead of a large final. Each test is a way for me to find out how to modify my lessons to help students understand better. After each test, students must make all test corrections and when they can show that they know all the material, they get full credit. My students know that no matter how long it takes, they can learn the material and get credit for it. They have told me that the tests help them to know what is important and what they still need to work on.

I no longer assign grades to homework, papers, and projects. Instead, I write comments directed toward improving the piece of work. Student work is not accepted until it meets my standard (which is documented on a rubric or assignment specification sheet) and then it gets full credit. Students are expected to revise their work until it meets an A standard. At first I tried to make the standards rigorous on every assignment and found that I was getting weighed down in paper. Then I moved to checking smaller assignments at the students' desks and assigning only one big project with rigorous standards every other week.

An atmosphere of self-improvement has become the accepted classroom norm. For some students, this is their first realization that they can get an A, and are expected to try to do so. I do not give the option of accepting a lower grade for work done and then quitting. I give time in class for revisions, so I can work with students that need the most re-teaching. I sometimes have students present their projects and give them private verbal feedback.

Teachers have asked, "Why not give grades and comments together?" This can negate the positive effects of comments. The work my son brings home from his elementary school is graded with a number score based on a rubric. Once he sees his score he is done looking at that paper. After reminding him that he can still learn from the paper if he looks at the teacher's comments, he has told me, "It doesn't matter any more; I already got graded on it." Many students internalize the grade as a judgment of themselves and come to accept this judgment. The motivation to continue working is minimal, especially if they have worked hard and received a D.

Is my system fair? What about the students that usually get high grades? Don't they deserve better grades than the students who take two or three tries to reach the same level? My answer is no. What they deserve is better education than they have received in the past. When I have a student who has met the basic standard on the first try, I push them to do more, to go further with the topic or theme because the focus is on learning, not on the grades. Since everyone else is working to improve, there is no resentment about this and students are not ostracized for having to revise work. Students who come in with lower skills in writing and reading have to work harder to reach the A standard. It does not matter to me who reaches high standards first, just that we all work to go as high as we can go.

Drawbacks exist. Some students do not feel the need to study for tests since they will have time to revise for full credit. I need to find a way to motivate these students. Time is an issue. I often have to assess the same assignment two or even three times for some students. Still, I find it rewarding to have students finally understand that density is a ratio of mass and volume instead of repeating the misconception that it is a measure of how heavy an object is.

IRENE'S PERSPECTIVE (USING OPEN-ENDED PROJECTS AND RUBRICS)

I have been teaching for 12 years, and have struggled with giving meaningful feedback to my students in a timely fashion. I generally start off the year writing lengthy comments on assignments and returning them soon after they are handed in. As the quarter progresses, however, students receive papers back from me weeks after turning in the assignment. In fact, sometimes so much time has passed that the students have forgotten doing the assignment.

I have found a few ways of making the feedback more meaningful—and more timely. I have limited the number of assignments on which I write detailed comments. I have students do many assignments "for completion only." Unlike the other assignments, I am able to get the notebooks back to the students in a day or two because I grade based on completion and thoroughness and whether the students self-corrected their work.

Long-term projects and lab reports have become easier to grade for two reasons. First, when the expectations are clearly written out in a rubric, students are able to self-assess. Second, I spend less time writing comments because I can just write brief comments next to the detailed criteria. For example, on lab reports, one assessment criteria pertained to error: "Experimental errors: Describe possible errors that were made and errors in the design of your experiment that may have affected your results." If a student just discussed errors he or she made I could just write, "What about errors you couldn't control?" Ultimately, my goal in giving feedback is for students to learn and improve their work from my comments. However, since I struggle to return their work quickly, I would like to move toward Acceptable/Not Acceptable grading for projects. Students learn by revising their work and retain the learned information longer if they have to revise their work. My students are also very grade driven. If the students are freed from grades and know that everyone can get 100%, they can focus on the learning and be more willing to help each other, rather than competing with each other. The freedom from grades should help to build self-esteem, especially in the low-achieving students.

I would like to develop open-ended projects that address the key concepts in the unit of study. Realistically, there would be one Acceptable/Not Acceptable project per grading period. The students would receive a detailed rubric explaining the project. The projects would consist of a model and/or demonstration with a written component as well. I would check off whether the students' work meets the standard; if the criteria were not met, I would write brief comments as to how the student could improve his or her work. I would write comments that are clear but don't tell the students "the answer." (If feedback is too specific, the students tend to mechanically go through the process of correcting but do not learn from their mistakes.) Furthermore, it is possible that the students won't not be able to self-correct. I would like students to apply what they learn to new situations, and I am not sure they would learn this skill with the revision process. Perhaps, if I introduce peer assessment before turning in the first draft, students can see their own mistakes by comparing their own work with another student's.

Acceptable/Not Acceptable projects can be helpful to the students but I have several concerns about that approach. One is coverage: If the students devote so much time to revising a project through multiple revision, will I be able to cover enough of the cur-

riculum to meet my satisfaction? A second issue involves manageability of grading. Since I have such large classes, providing individual feedback is difficult. If a rubric is clear, detailed, and student friendly, students will be able to assess their work before they turn work in. I can also organize project time so that I can give oral feedback while they are engaged in their work. Group work would also make verbal feedback more manageable. I also would need to keep other concerns in mind: Will students lose interest or become frustrated when not successful in the revision process? Will they just copy another student's acceptable answer? And will my moving on to new topics while some students are still revising a project contribute to a disconnected experience?

DISCUSSION: LEARNING FROM PRACTICE

One of the guiding premises of CAPITAL was that teachers' practices and changes are connected to their priorities, beliefs, and specific classroom situations. As teachers explored different types and forms of feedback that were useful for their students and logistically manageable, their views and practices evolved (into a variety of assessment scenarios).

Despite the variations in the five teachers' practices and perspectives, common themes emerged. There were inextricable connections among assessment, curriculum, teaching, and learning; even the subtlest shifts in the forms and types of feedback had effects in other areas. For example, in Anthony's class, "coverage of material" gave way to 'increased student involvement in assessment" and "increased student understanding of the criteria for which they were accountable." As Joe grappled with issues of fairness, he confronted questions of equity, the purposes of assignments, and even the purpose of schooling. Irene highlighted some of the trade-offs endemic to changing one's practice—juggling competing priorities, including the major struggle of managing paperwork.

Feedback came to mean not just giving student grades, marks, scores, or comments; feedback eventually dealt with content, quality, timing, and form. Appropriate feedback was critical for helping students better understand the content and confront misconceptions. Feedback also was necessary to help students to understand more clearly what counted as understanding and to demonstrate their understanding of science. Jen aptly pointed out that it was often easy to confuse understanding with the articulation of understanding. Feedback from student work informed the teachers about adjustments they needed to make to their instruction. Quality feedback supported student learning and teacher learning, and helped teachers and students learn from each other.

Another common theme was the integral roles that students played in the assessment process, on numerous levels. Students were not only the recipients of feedback, they also contributed feedback to their teachers and peers. In this way, students developed a better sense of the meanings underlying assessment criteria, and the relevant conventions of communicating and knowing in the science classroom. They needed to be engaged directly in the process, so that they could critically examine feedback, rather than just accepting it. As Caleb's remarks about his policy of posting grades suggested, students interacted with assessment, often in ways not predicted by their teachers.

As the CAPITAL teachers delved deeper into their practices—reflecting, sharing, trying new things—they met with and embraced increasing complexity. They developed more nuanced understandings of assessment, their students, themselves, and how practices developed and played out in their classrooms. Over time, their responses and reflections became more sophisticated and the teachers generated more questions than answers.

LINKS TO THE NATIONAL SCIENCE EDUCATION STANDARDS

The National Science Education Standards (NRC 1996) state that assessment should meet the full range of goals for science education. Both the Assessment and Teaching Standards make a compelling case for the tight coupling of assessment, teaching, and learning. The breadth of purposes for assessment is captured in Teaching Standard C. The Teaching and Professional Development Standards also highlight the benefit of discussing practice collaboratively with colleagues.

REFERENCES

Atkin, J. M. 1992. Teaching as research: An essay. *Teaching and Teacher Education* 8 (4): 381–390.

Black, P. J., and D. Wiliam. 1998a. Assessment and classroom learning. *Assessment in Education* 5 (1): 7–74.

Black, P. J., and D. Wiliam. 1998b. Inside the black box: Raising standards through formative assessment. *Phi Delta Kappan* 80 (2): 130–148.

National Research Council (NRC). 1996. *National science education standards*. Washington, DC: National Academy Press.

Ramprasand, A. 1983. On the definition of feedback. *Behavioural Science* 28 (1): 4–13.

Schwab, J. J. 1969. *College curriculum and student protest*. Chicago: University of Chicago Press.

Wenger, E. 1998. *Communities of practice*. Cambridge, UK: Cambridge University Press.

RESOURCES

Additional information about the Classroom Assessment Project to Improve Teaching and Learning (CAPITAL) (NSF Grant REC-9909370) can be obtained by contacting Janet Coffey at *jecoffey@umd.edu* or Savitha Moorthy at *smoorthy@stanford.edu*. Mailing Address: CAPITAL Project, School of Education, CERAS 407, 520 Galvez Mall, Stanford, CA 94305.

AUTHOR AFFLIATIONS

Caleb Cheung and **Anthony Cody** teach middle school science in the Oakland Unified School District in Oakland, California.

Irene Hirota and **Joe Rubin** teach middle school science in the San Francisco Unified School District in San Francisco, California.

Jennifer Slavin teaches middle school science at Eastide Preparatory School in East Palo Alto, California.

Janet Coffey and **Savitha Moorthy** were research assistants with the Classroom Assessment Project to Improve Teaching and Learning at Stanford University. Coffey, now an assistant professor of curriculum and instruction in the College of Education at the University of Maryland, formerly taught middle school science. She received her PhD in science education in 2002. Moorthy, who recently completed her doctoral work at Stanford University, previously taught English language arts.

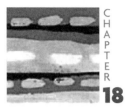

mind mapping as a flexible assessment tool

Karen Goodnough and Robin Long

INTRODUCTION

Assessment of student learning in science is carried out for a variety of purposes ranging from assigning student grades to monitoring the impact of educational policy. Calls for education reform in science (AAAS 1989; NRC 1996) have promoted a notion of scientific literacy that requires students to become familiar with the content knowledge of science and with ways to engage in scientific inquiry and problem solving; develop an understanding of the history and development of science and the nature and methods of science; and acquire an understanding of the complex interplay among science, technology, society, and the environment. To support the goal of scientific literacy, students will need opportunities to experience learning and assessment in a variety of ways that consider students' individual learning needs, interests, abilities, and backgrounds.

With the growing diversity in the student population and the need to provide students with a variety of active, student-centered learning and assessment experiences, it is important for teachers to be comfortable with and employ a range of assessment approaches. This chapter focuses on a collaborative, classroom-based research project conducted by a teacher researcher and a university researcher exploring mind mapping. The mind map is a visual tool (Buzan 1983) to improve note taking, foster creativity, organize thinking, and develop ideas and concepts. Because instruction, learning, and assessment are inextricably linked, we focused on the efficacy of the mind-mapping tool as a teaching, learning, and assessment tool in a grade 6 science classroom. Little systematic research exists about the effectiveness of mind mapping as a learning and assessment tool. This study provided preliminary evidence for the adoption of mind mapping as a flexible assessment tool to foster student learning in science and to guide curriculum planning and classroom practice.

BACKGROUND

Graphic organizers come in many varieties—for example, Venn diagrams, concept webs, flow charts, and concept maps—and are used for a range of purposes. Derived from Ausubel's work (e.g., Ausubel 1960) on advanced organizers (or introductory prose passages), advanced organizers were later modified and became structured overviews, incorporating both graphical information and text (before and after instruction). Eventually, the structured overview was replaced by the "graphic organizers." Few graphic organizers have been studied as systematically as concept maps (Novak and Gowin 1984; Mintzes, Wandersee, and Novak 1998, 2000). Research on concept maps has shown many positive results for science teaching, learning, and research (Novak and Musconda 1991; Horton et al. 1993). By contrast, mind mapping has little empirical research supporting its use in classrooms.

Guidelines were adopted for creating mind maps (Buzan 1983), which were summarized by Wycoff (1991, p. 43): "A central focus or graphic representation of the problem is placed in the center of a page; ideas are allowed to flow freely without judgment; key words are used to represent ideas; one key word is printed per line; key words are connected to the central focus with lines; color is used to highlight and emphasize ideas; and images and symbols are used to highlight ideas and stimulate the mind to make connections." A mind map of how the mind-mapping tool was conceptualized and applied in the study discussed here is shown in Figure 1.

A RATIONALE FOR MIND MAPPING

Classroom assessment has often been viewed as something that is completed at the end of a unit or sequence of study to judge the overall performances of students. This approach, referred to as summative assessment, provides little feedback for teachers or students about how to improve the learning process. In contrast, formative assessment (an ongoing part of instruction) provides the teacher with more immediate feedback about students' abilities, interests, and understandings. The teacher can modify instruction accordingly, especially if students are not grasping ideas and concepts, or need more time to develop a skill.

Many science teachers acknowledge that assessment is a dynamic process that involves continual interaction among teachers, learners, resources, and assessment instruments (Knutton 1994). They employ a variety of assessment tools over time to obtain a more accurate and fairer picture of student achievement, abilities, skills, and dispositions. Thus, the more assessment tools teachers have at their disposal, the more opportunities and ways they have to get feedback from students and to plan their science instruction. There are many kinds of formative assessment—journals, notebooks, interviews, observations, products, and self-as-

FIGURE 1.

A Mind Map by the Authors About the Project

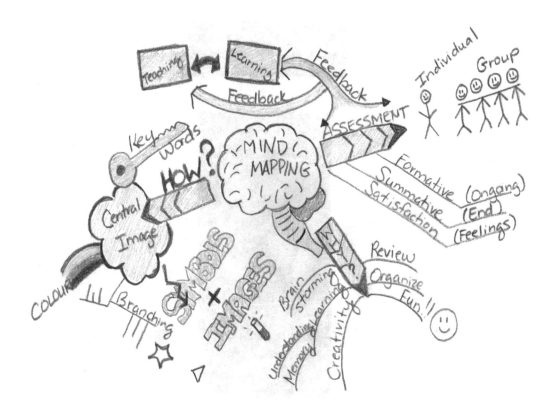

sessments. The kind used depends on factors such as a teacher's goals and objectives and students' learning styles. Visual tools, such as mind mapping, provide teachers and learners with another option for exploring the developing understanding of ideas and concepts and prior knowledge brought to a learning situation. Mind maps can be used for both formative and summative assessment.

The use of mind mapping is also consistent with current theories of learning. For example, Gardner's theory (1999) of multiple intelligences (a pluralistic conception of intelligence posited on the existence of eight distinct intelligences—verbal-linguistic, logical-mathematical, visual-spatial, musical-rhythmic, bodily-kinesthetic, interpersonal, intrapersonal, and naturalist—provides a helpful, commonsense framework for teachers.

Another framework, constructivism, places the individual at the center of knowledge creation. Individuals construct their own meaning and understanding of the world, and generate knowledge through the interaction of what they know and believe and the ideas and experiences they encounter. Those who adhere to social constructivist views of learning emphasize the social creation of knowledge. In this perspective, the creation of scientific knowledge occurs through critical, public dialogue within a global community of scientists. Brooks and Brooks (1993) identified five basic tenets of constructivism, including seeking and valuing students' points of view, challenging students beliefs and ideas, making the curriculum relevant, structuring learning that focuses on big ideas in the curriculum, and assessing students in the context of classroom events in a variety of ways.

Mind mapping provides opportunities for students to construct meaning in both individual and group contexts in classrooms. Mind mapping and other visual tools, if used appropriately, have the potential to support the tenets of constructivism.

THE MIND-MAPPING PROJECT

The collaborators in this project were Karen, a university researcher, and Robin, a veteran middle school science teacher with 15 years experience. We started our planning and research in fall 2000, and gradually introduced the mind-mapping skill to students (Buzan 1983). By the end of the fall term, we felt the students were competent mind mappers and ready to use the mind-mapping skill in diverse, systematic ways in an online curriculum unit called "Blue Ice."

Robin's school was a small private, independent school in upstate New York that provided an informal learning environment. The population was largely suburban; 10%–15% was minority students. Mind mapping was used in two of Robin's grade 6 classes (formal data were collected only from one group, consisting of nine females and seven males who were heterogeneous in terms of academic ability). Students had used a range of visual tools in the past, but none had used mind mapping prior to the start of the project.

RESEARCH QUESTIONS, METHOD, AND TREATMENT

These research questions guided the study: (a) In what ways is mind mapping used as an instruction and assessment tool when students are learning and doing science? (b) How do students perceive mind mapping after using it for an extended period of time? and (c) How does mind mapping assist students in learning science?

In this interpretive case study (Merriam 1998; Stake 1994), we adopted a collaborative inquiry approach (engaging in ongoing cycles of planning, action, and reflection) to better inform and guide the unfolding research. Data were collected using several methods and sources for triangulation over a period of time. Member checking occurred on an ongoing basis, as we shared, analyzed, and interpreted data collaboratively. The project took place over a 10-month span, from September 2000 to June 2001. Data collection methods and sources were the following:

Semi-structured interviews. At the end of the study, each student participated in a 20–30 minute semi-structured interview discussing his or her understanding of mind mapping as a learning tool. The interviews were audiotaped and transcribed; careful notes were taken after each interview.

Field notes. Karen visited the students' classroom at least three times per week for instructional periods lasting up to 90 minutes. During these visits, she was a participant-observer (acting as a resource, working with small groups of students) and co-teaching with Robin. Karen took copious notes both while observing and shortly after the observation period. Robin also kept observational notes; both sets of notes served as a means to cross-check observation data and corroborate interpretations of these data.

Open-ended questionnaire. At the end of the study, students responded in writing to eight open-ended questions about their understandings and perceptions of mind mapping. These questions were also used as part of the interview schedule of questions. Students had opportunities to elaborate on their written responses. The researchers compared different types of data for the convergence of themes.

Documents. Student-generated documents—samples of writing, mind maps, models, and other products—provided another means for triangulation, or viewing the research situation from several perspectives. Lesson plans and other materials produced by Robin also became data sources, enriching the analysis and interpretation of data.

Data analysis and interpretation began early in the study. Most data were in the form of free-flowing texts consisting of short segments. These units of text were assigned labels (Coffey and Atkinson 1996) to allow intensive data analysis later. At the second level of coding, labels were used to establish categories. The researchers identified concepts from various sources to refute or support broad themes that guided data collection and interpretation.

Formal data collection occurred when Robin and Karen implemented an eight-week online unit, "Blue Ice" (Online Class 2000). Our primary learning goals were to help students develop an understanding of the interrelationship between animals and their environment, the basic requirements of habitats, and the impact of human activity on ecosystems. The program was organized into eight units having two distinct threads—food webs and weather/climate change—that could be mixed or matched to meet the needs of individual classrooms. We selected the food web thread for this study. Several online components (background resources for the unit, sharing ideas with other schools, and ask-a-scientist) were part of the unit; most of the learning activities occurred off-line.

Before using mind mapping with students in the "Blue Ice" unit, we spent a considerable amount of time helping students develop an understanding of mind mapping and how to create mind maps. We began from a global perspective by creating a large mind map on the blackboard for which students provided ideas and information about their school. The map was intended to provide a visitor with information about the school as well as give students an introduction to mind mapping. This activity was followed by an explicit presentation on the rules for mind mapping and a class brainstorming session about ways that mind mapping might be used to learn science. Then, Robin spent time exploring symbols, a critical component of mind mapping.

Throughout the semester, Robin used mind mapping in her teaching. At times, she placed a mind map on an overhead as an introduction to a new topic. At other times, she created a mind map with the class to review a lesson or series of lessons or to brainstorm ideas. On other occasions, she asked students to add color, symbols, and key words to partially completed mind maps. Eventually, students created their own maps. In addition to mind maps, a range of assessment and learning activities was used throughout the unit, including paper-and-pencil tests, journal entries, projects, presentations, interviews, informal observations and discussions, essay writing, and models.

FORMATIVE ASSESSMENT

Robin used mind mapping as an assessment tool to provide feedback about her teaching and to offer students ongoing feedback about their learning. One learning activity required students to create a very large class mind map (5' by 12') to reveal students' understandings of the characteristics of Antarctic animals, food chains, and webs and the nature of the Antarctic environment. Students were organized into teams of three or four; each team researched information on two organisms from the Antarctic food web and constructed parts of the overall class mind map (located in a hallway outside the classroom). Before starting this whole-class collaborative project, Robin reviewed important principles for fostering effective teamwork. Each team then developed a mind map that outlined their conceptions of effective teamwork. An informal, online quiz about team skills followed. In this way, mind mapping served two purposes: It helped focus students' attention on the nature of effective teamwork and it provided feedback about students' notions of what constituted effective collaboration. Robin was able to introduce and reinforce important ideas that students had not considered.

To promote individual accountability, each student created an individual mind map to ensure that his or her developing ideas about food webs and habitats were accurate. Periodically, students were supplied with a list of concepts and asked to create an individual mind map showing understanding and connectedness of the concepts. This nontraditional format gave students another way to express what they were learning. In several instances, students who struggled with a written assignment expressed their ideas in clear and accurate ways by using mind mapping.

Student self-assessment was an integral part of the unit. Students made regular entries in their journals about how their team functioned and how each person contributed to the overall functioning of the team. Robin provided guiding questions such as, "Describe a problem your team had and how it was handled. If it was not handled well, what suggestions do you now have that may have made the situation better?" She encouraged students to share their thinking and ideas through written means and their developing ideas and feelings through mind maps.

RESULTS

SUMMATIVE ASSESSMENT

After using a range of formative assessments, Robin assessed individual student learning in terms of the instructional objectives and goals at the end of the unit of study. Some components of the formative assessment were used to determine an overall grade. One of the main components of summative assessment included the creation of a mind map that integrated all concepts and ideas in the unit. The concepts included in the mind map and a rubric were provided to guide creation of the maps. The rubric focused on the scientific accuracy of connections among ideas, the development of ideas through branching, and presentation (following mind-mapping rules). The rubric was used to individually assess student work and compare assigned grades. Students scored an overall 4 (above average) or 5 (exemplary) on the mind maps.

Although students had a solid understanding of the ideas and concepts related to food webs and Antarctica, they still struggled with the guidelines for creating mind maps. The guideline that presented the biggest challenge was using one key word per branch to express or represent an idea (a challenge that is inherently part of designing mind maps). Instead, students used strings of text to explain an idea. The creation of mind maps demanded higher-level thinking and required students to represent their understandings of concepts and relationships using text, graphics, and symbols.

SATISFACTION ASSESSMENT

"Satisfaction assessment" means finding out how students feel about their learning (Trice 2000). Feelings about learning may not indicate, however, how much learning is actually occurring. We collected satisfaction data informally (through observations and field notes) and more formally through open-ended student questionnaires and individual student interviews. Several themes emerged when analyzing this information:

* Students reported that mind mapping was motivating and fun.
* Students expressed preferences for individual versus group mind mapping.
* The majority of students stated that mind mapping enhanced their learning in a variety of ways.

At least two-thirds of the class attributed the fun part of mind mapping to the latitude for expressing their creativity through color, symbols, and individual design.

All students believed that mind mapping was a valuable tool for enhancing their understanding of science by fostering motivation, enhancing memory, deepening their reflections about ideas and how they were learning, and improving organization of concepts. Ten percent of students stated that they preferred other approaches to learning. Upon further probing, one student stated that he did not like group mind mapping because it required considerable negotiation to create the mind map. Another student, when given a choice, told us she preferred to learn through reading and writing to express her ideas and to learn new ideas.

CONCLUSION AND IMPLICATIONS FOR PRACTICE

Based on evidence collected from this project and our opinion, mind mapping has the potential to be a flexible assessment tool to ascertain students' developing ideas about scientific concepts and ideas. Mind mapping can be used to explore prior knowledge, help consolidate students' developing understanding throughout a unit of study, and gather summative information about students' overall performances related to specified learning outcomes. Equally important is the need to offer those learners who have unique, diverse learning needs a range of assessment tools that can provide both them and their teachers with feedback about learning. By designing assessments that target these varieties of intelligences (Gardner 1999), teachers can address different learning styles in science classrooms. In addition, teachers can integrate mind mapping with other instructional approaches (such as cooperative learning) to enrich students' classroom experiences.

Mind mapping provides a unique window into the development of students' understandings in science, while simultaneously providing a venue for students to express their creativity and individuality. More research is needed to explore the use of mind mapping at other grade levels. We invite science teachers to engage in classroom inquiry by exploring the efficacy of mind mapping as both an assessment tool and a learning tool. In our opinion, mind mapping can be used as part of a comprehensive repertoire of constructivist-based learning and assessment tools to improve student learning.

LINKS TO THE NATIONAL SCIENCE EDUCATION STANDARDS

The assessment practices described in this chapter target several aspects of the National Science Education Standards (NRC 1996). In Standard A, assessments are expected to be "consistent with the decisions they are designed to inform." In this study, the mind-mapping tool was used for three specific purposes—to motivate students, to ascertain their developing understanding of scientific concepts and principles, and to allow students opportunities to develop and consolidate their understandings of scientific concepts and principles. Standards B and C were also addressed: "Equal attention is given to the assessment of opportunity to learn and to assessment of student achievement" (Standard B); students were given "adequate opportunity to demonstrate their achievements" through the use of mind mapping for different learning and assessment tasks (Standard C).

REFERENCES

American Association for the Advancement of Science (AAAS). 1989. *Science for all Americans: Project 2061.* New York: Oxford University Press.

Ausubel, D. P. 1960. The use of advance organizers in the learning and retention of meaningful verbal material. *Journal of Educational Psychology* 51: 267–272.

Brooks, G., and M. G. Brooks. 1993. *In search of understanding: The case for constructivist classrooms.* Alexandria, VA: Association for Supervision and Curriculum Development.

Buzan, T. 1983. *Use both sides of your brain.* New York: E.P. Dutton.

Coffey, A., and P. Atkinson. 1996. *Making sense of qualitative data: Complementary research strategies.* Thousand Oaks: Sage.

Gardner, H. 1999. *Intelligences reframed: Multiple intelligences for the 21st century.* New York: Basic Books.

Horton, P. B., A. A. Conney, M. Gallo, A. L. Woods, G. J. Senn, and D. Hamelin. 1993. An investigation of the effectiveness of concept mapping as an instructional tool. *Science Education* 77 (1): 95–111.

Knutton, S. 1994. Assessing children's learning in science. In *Secondary science: Contemporary issues and practical approaches,* ed. J. Wellington, 72–93. New York: Routledge.

Merriam, S. B. 1998. *Qualitative research and case study applications in education.* San Francisco: Jossey-Bass.

Mintzes, J., J. Wandersee, and J. Novak. 1998. *Teaching science for understanding.* San Diego, CA: Academic Press.

Mintzes, J., J. Wandersee, and J. Novak. 2000. *Assessing science understanding.* San Diego, CA: Academic Press.

National Research Council (NRC).1996. *National science education standards.* Washington, DC: National Academy Press.

Novak, J. D., and D. B. Gowin. 1984. *Learning how to learn.* New York: Cambridge University Press.

Novak, J. D. and D. Musonda. 1991. A twelve-year longitudinal study of science concept learning. *American Educational Research Journal* 28 (1): 117–153.

Online Class. 2000. *Blue ice.* TBT International. Available online *at www.onlineclass.com/BI/.*

Stake, J. E. 1994. Development and validation of the six-factor self-concept scale for adults. *Educational and Psychological Measurement* 54 (1): 56–72.

Trice, A. D. 2000. *A handbook of classroom assessment.* New York: Addison-Wesley Longman.

Wycoff, J. 1991. *Mindmapping: Your personal guide to exploring creativity and problem-solving.* New York: Berkley Publishing Group.

RESOURCES

Hyerle, D. 1996. *Visual tools for constructing knowledge.* Alexandria, VA: Association for Supervision and Curriculum Development.

This comprehensive overview offers a rationale for using visual tools, explores the three types of visual tools (brainstorming webs, task-specific organizers, and thinking-process maps), and discusses mind mapping and its purposes.

Inspiration Software, Inc. 2002. *Inspiration* [Computer software]. Retrieved from *www.inspiration.com.*

Inspiration is a computer software program that allows the user to create a variety of graphic organizers including concepts maps, webs, flowcharts, mind maps, and other visual representations. A free 30-day version of the software can be downloaded or a copy of the software can be ordered online.

Margulies, N. 1991. *Mapping inner space: Learning and teaching mind mapping.* Tucson, AZ: Zephyr Press. This book provides an excellent introduction to mind mapping, leading the reader through the process of mind mapping and providing detailed guidance about how to develop students' mind-mapping skills (book no longer in print).

AUTHOR AFFILIATIONS

Karen Goodnough is an associate professor at Memorial University of Newfoundland, St. John's, Newfoundland, Canada. Her teaching and research interests include science teaching and learning, teacher development, action research, and problem-based learning.

Robin Long has taught middle school science for 15 years in a small, independent, private school in upstate New York.

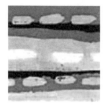

index

Page numbers printed in **bold** type reference information contained in figures, illustrations, or tables.